AUTHOR CO-CITATION ANALYSIS
USING CUSTOM BIBLIOGRAPHIC DATABASES

AUTHOR CO-CITATION ANALYSIS USING CUSTOM BIBLIOGRAPHIC DATABASES
An Introduction to the SAS Approach

Sean B. Eom

The Edwin Mellen Press
Lewiston•Queenston•Lampeter

Library of Congress Cataloging-in-Publication Data

Eom, Sean B.
   Author co-citation analysis using custom bibliographic databases : an introduction to the SAS approach / Sean B. Eom.
     p. cm.
   Includes bibliographical references and indexes.
   ISBN 0-7734-6713-0
   1. Bibliometrics. 2. Bibliography--Databases. 3. Multivariate analysis. 4. SAS (Computer file). 5. Research--Statistical methods--Data processing. I. Title.

Z669.8.E56 2003
010.'727--dc21

2003051372

A CIP catalog record for this book is available from the British Library

Copyright © 2003 Sean B. Eom

All rights reserved. For information contact

      The Edwin Mellen Press   The Edwin Mellen Press
      Box 450   Box 67
      Lewiston, New York   Queenston, Ontario
      USA 14092-0450   CANADA L0S 1L0

      The Edwin Mellen Press, Ltd.
      Lampeter, Ceredigion, Wales
      UNITED KINGDOM SA48 8LT

Printed in the United States of America

This book is dedicated the following individuals who have shaped who I am today.

To My Mother, Jong Ye Won
She planted a seed on a desert. Her tears have made the seed grow to bear a fruit.

To Dr. Sang M. Lee
Dr. Lee was my doctoral advisor. He has given me extraordinarily strong drive toward the goal. This book could have not been written without his influence on my life.

These two individuals have driven me at all times, wherever I go. They have given me a spirit that can overcome any earthly adversity.

Contents

INTRODUCTION 1
PRIMARY OBJECTIVE OF THIS BOOK 2
ORGANIZATION OF THE BOOK 6

CHAPTER 1 BASIC CONCEPTS OF BIBLIOMETRICS AND AUTHOR COCITATION ANALYSIS 7

WHAT IS BIBLIOMETRICS (STATISTICAL BIBLIOGRAPHY)? 7
   Citation Analysis 8
   Author Cocitation Analysis 9
ASSUMPTIONS OF AUTHOR COCITATION ANALYSIS 11
PURPOSES AND BENEFITS OF AUTHOR COCITATION ANALYSIS 12
LIMITATIONS /CRITICISMS OF AUTHOR COCITATION ANALYSIS 13

CHAPTER 2 BUILDING THE BIBLIOGRAPHIC DATABASES 15

DATABASE DESIGN 16
   The Structure of the Bibliographic File 16
POPULATING DATABASE 18
   Selection Criteria For Creating The Citing and cited Article Database .. 18
   Referenced (Cited) Article Database 18

CHAPTER 3 OVERVIEW OF AUTHOR COCITATION ANALYSIS STEP 23

SELECTION OF AUTHORS/COCITATION THRESHOLDS 23
   Subjective (Top-down) Approach 24
   Objective (Bottom-up) Approach 26
      The loose screening to produce a preliminary list 26
      Finalizing the list of author via detailed analysis 27
   Ad hoc Criteria for Further Screening of Authors 28
GENERATION OF COCITED AUTHOR COUNTS 29
   The Cocitation Count Generation System 30
   Raw Cocited Author Counts 35
   Creation of Raw Cocitation Matrix 36
   Conversion of Raw Cocitation Matrix 36
      Adjusting diagonal cell values 36
      Transposing adjusted diagonal cell value cocitation matrix 38
A HYPOTHETICAL CASE OF A SIMPLIFIED COCITATION MATRIX 38
      Conversion of raw cocitation matrix with adjusted values in diagonal values 39

CHAPTER 4 THE FACTOR PROCEDURE 41

OVERVIEW OF INPUT, PROCEDURES, AND OUTPUTS 42

DEFINITION OF FACTOR ANALYSIS.. 45
GETTING THE DATA INTO A SAS DATA SET 46
  Manual Data Inputs without In-File Statements or Permanent Data Sets... 46
    The DATA statement... 48
    The INPUT statement 48
    The CARDS statement... 49
  Data Inputs using Permanent SAS Data Sets.. 50
    The INFILE statement 51
    Retrieval from permanent SAS data sets 52
PROC FACTOR STATEMENT 54
  Data Options (Data=) 54
  Factor Extraction Options (Method=) 54
  Specifying Number of Factors Options (MINEIGEN=, NFACTORS=)........ 55
    MINEIGEN.. 55
    NFACT.. 55
  Specifying Rotation Method Option 56
  Displaying Output Options (Scree) 57
EXAMPLES OF ACA PROC FACTOR PROGRAMS 58
  Manual Data Inputs With Author's Names 58
  With an Embedded In-File Statement... 59
PROCESSING THE INPUTS 60
INTERPRETING THE FACTOR PROCEDURE OUTPUTS................................... 63
  Initial Factor Method: Principal Components... 63
    Optimal number of factors 66
  Prerotation Method: Varimax... 70
    Rotated factor pattern.. 71
  Rotation Method: Promax... 73
    Interfactor correlations.. 73
    Factor structure correlations... 75

CHAPTER 5 THE CLUSTER PROCEDURE 77
GENERATION OF A DISTANCE MATRIX USING XMACRO.SAS AND DISTNEW.SAS. 78
  The %INCLUDE Statement 78
  %Distance Arguments.. 80
    Specify the input data set 80
    Print the name of all authors in the outputs...................................... 80
    List additional options... 81
    Specify the method for computing distance measures..................... 81
    Specify the variables in the input data sets 82
    Specify the output data set to be produced....................................... 82
CREATING THE PERMANENT DISTANCE MATRIX... 84
PROC CLUSTER STATEMENT... 86
  PROC CLUSTER METHOD=name <options>; 87
    Specify the input data set 87
    Specify clustering methods 87

Control display of the cluster history 88
ID variables 89
INTERPRETING RESULTS OF CLUSTER ANALYSIS 92
Data Set WORK.DIST 92
Cluster History 93
Selecting the Number of Clusters 94
   Plot of the pseudo F statistic against number of clusters 94
   Plot of the pseudo t^2 statistic against number of clusters 96
Proc Tree Graph Output 97

CHAPTER 6 MUTIDIMENSIONAL SCALING 101

THE MDS PROCEDURE 102
Similarity/Proximity Measures 102
Proc MDS Statement 105
   Specify the input data set 106
   Specify the out data set 106
   Specify the type of the dimension coefficients 106
   Specify the conditionality of the data 107
   Specify the measurement level of the data and the type of transformations 107
   Specify the number of dimensions 107
   Specify a predetermined transformation 107
   Miscellaneous 108
   ID statement 108
   Control displayed output 108
MDS Procedure Output 110
   Iteration history and convergence status 110
THE PLOT PROCEDURE 113
PROC PLOT Statement 113
Proc Plot < option(s)> 115
   Specify the input data set 115
   Control the appearance of the plot 115
PLOT plot-request(s) </ option(s)> 117
   Control axes by specifying the tick-mark values 117
   Use a box around the plot 117
PLOT Procedure Output 122
THE G3D PROCEDURE 128
Syntax of the G3D Procedure 128
   Generating a simple scatter plot 128
   Modifying plots with SCATTER options 129
CREATING THE ANNOTATE DATA SET USING THE DATA STEP 131
SAS Program to Create the Annotate Data Set (MYSAVE.ACAANNO) 131
Understanding the SAS Program for Creating Annotate Data Set 133
   Data <Data Set Name> 133
   Set 133

Length 134
Retain 134
Function 136
Coordinate-system 136
Assignment 136
PROC G3D with the Annotate =Option *138*

CHAPTER 7 AN ACA STUDY IN AN INFORMATION SYSTEMS AREA: THE INTELLECTUAL STRUCTURE OF DECISION SUPPORT SYSTEMS RESEARCH (1969-1990) **145**

INTRODUCTION 146
DATA 149
RESEARCH METHODOLOGY 149
  Selection Of Author *150*
  Compilation Of Cocitation Frequency Matrix *150*
  Data Analysis *151*
RESULTS OF FACTOR ANALYSIS 152
  Reference (Contributing) Disciplines *157*
    Organizational sciences 157
    Multiple criteria decision making 157
    Group decision making 158
  DSS Research Subspecialties *158*
    Foundations 158
    GDSS 160
    Model/Data management 162
    User interface/Individual differences 163
    DSS implementation 165
RESULTS OF CLUSTER ANALYSIS AND MULTIDIMENSIONAL SCALING 167
MAJOR CONTRIBUTIONS OF ORGANIZATIONAL SCIENCE TO THE DEVELOPMENT OF DSS RESEARCH SUBSPECIALTIES 175
  Contributions Of Organizational Science To The Foundations Of Decision Support Systems *176*
  Contributions Of Organizational Science To The User Interface/Individual Differences *178*
  Contributions Of Organizational Science To Model Management *179*
CONCLUSIONS 180

REFERENCES **183**
SUBJECT INDEX **191**
NAME INDEX **195**

Illustrations

Figures

Figure 1.1 Can You See the Trunk, Branches, and Roots of the Tree? 10
Figure 1.2 ACA: A Tool for Digging Up the Roots, Examining the Trunk, and Identifying the Branches 10
Figure 2.1 An Example of Cited Reference Record 17
Figure 2.2 Citation Pattern 22
Figure 3.1 Author Cocitation Analysis Steps 25
Figure 3.2 Main Menu of Author Cocitation Frequency Generation Systems 31
Figure 3.3 Retrieval of Cocitation Frequency between Two Authors 32
Figure 3.4 Displaying Cocitation Frequency of a File 33
Figure 3.5 Cocitation Frequency Matrix Preparation Steps in ACA 34
Figure 4.1 Factor Procedure in ACA 44
Figure 4.2 The SAS Display Manager System Screen (Release 8.01) 62
Figure 4.3 The SAS Display Manager System Screen with Factor Analysis Data and Corresponding Outputs (Release 8.01) 62
Figure 4.4 Scree Plot of Eigenvalues 69
Figure 5.1 The Cluster Procedure in ACA 79
Figure 5.2 Plot of Pseudo F Statistic against Number of Clusters 95
Figure 5.3 Plot of Pseudo F Statistic against Number of Clusters using Birth and Death Rates in 74 Countries 95
Figure 5.4 Plot of Pseudo t^2 Statistic against Number of Clusters 97
Figure 5.5 Plot of Pseudo t^2 Statistic against Number of Clusters using Birth and Death Rates in 74 Countries 97
Figure 5.6 Dendrogram (Tree Graph) Depicting Cluster Structure and Joining Sequences 98
Figure 6.1 MDS/PLOT/G3D Procedures in ACA 104
Figure 6.2 Plot of Dim1* Dim2$Author 114
Figure 6.3 PROC PLOT Statement with the HAXIS, YAXIS Options 118
Figure 6.4 PROC PLOT Statement with the BOX, VTOH Options 118
Figure 6.5 Plot of Dim1*Dim2$Author with the HAXIS, YAXIS Options 119
Figure 6.6 Plot of Dim1*Dim2$Author with VTOH = 1 120
Figure 6.7 Plot of Dim1*Dim2$Author with VTOH=3 121
Figure 6.8 Plot of Dim1*Dim3 123
Figure 6.9 Plot of Dim2*Dim3 124
Figure 6.10 A Simple G3D Scatter Plot 129
Figure 6.11 A Scatter Plot with Options 130
Figure 6.12 Scatter (Dim1*Dim2) = Dim3 143
Figure 7.1 Major Factor Correlation Network 156

Figure 7.2 Dendrogram Depicting Cluster Structure and Joining Sequences (1970-1990) 169
Figure 7.3 Three-Dimensional MDS Map (1970-1990) 170
Figure 7.4 Two-Dimensional MDS Map (1970-1990) 174

Tables

Table 2.1 The Structure of Cited Reference File 16
Table 2.2 Citation Behavior 20
Table 3.1 Part of The Cocitation Count Generation System Output 35
Table 3.2 Diagonal Cell Value Adjusted Cocitation Matrix (1970-1990) 37
Table 3.3 Transposed Cocitation Matrix (1970-1990) 38
Table 3.4 Reference of Citing Papers 39
Table 3.5 Sample Cocitation Matrix 39
Table 4.1 Creating a Temporary Data Set with Variable Number 47
Table 4.2 Creating a Temporary Data Set with Author Names in the Input Statement 50
Table 4.3 Creating a Temporary Data Set with Line Pointer Control Statements 50
Table 4.4 SAS Data File (9varname.dat) Stored on c:\wp\books\aca\sasdata\ 52
Table 4.5 SAS Data File (9varnoname.dat) Stored on C:\wp\books\aca\sasdata\ 52
Table 4.6 SAS Program Creating a Permanent Dataset 53
Table 4.7 Permanent File (mysave.acaname) 53
Table 4.8 PROC FACTOR SAS Program with Author's Names 58
Table 4.9 PROC FACTOR SAS Program using Author's Name with an INFILE Statement 59
Table 4.10 Factor Procedure using a Permanent data Set 59
Table 4.11 Initial Factor Method: Principal Components 64
Table 4.12 A Matrix of Factor Loadings with Eigenvalue and Communality 66
Table 4.13 Rotated Factor Pattern (Varimax) 72
Table 4.14 Interfactor Correlations (1970-1990) 74
Table 4.15 Factor Structure Correlations 75
Table 5.1 Creating Permanent Distance Data Set with Card Input 85
Table 5.2 Creating Permanent Distance Data Set with INFILE Statement 85
Table 5.3 PROC CLUSTER SAS Statement with a Permanent Data Set 89
Table 5.4 Options in the PROC CLUSTER Statement 90
Table 5.5 PROC CLUSTER Statement 91
Table 5.6 Data Set Work.DIST 92
Table 5.7 Cluster History 93
Table 6.1 PROC MDS with a Permanent Data Set (Mysave.dist) 105
Table 6.2 MDS Procedure Output 110
Table 6.3 Iteration History and Convergence Status 111
Table 6.4 Configuration 112
Table 6.5 Data Set MYSAVE.COORD 112
Table 6.6 PLOT *plot-request(s)* </ *option(s)*>; 116
Table 6.7 PROC MDS SAS STATEMENT 125
Table 6.8 PROC MDS with a Permanent Data Set (Mysave.dist) 127
Table 6.9 Creating an Annotate Data Set 131

Table 6.10 An Annotate Data Set 132
Table 6.11 The Value of The Coordinate System... 137
Table 6.12 PROC G3D Procedure with Temporary Data Sets.......................... 139
Table 6.13 PROC G3D Statement with Permanent Data Sets........................... 141
Table 7.1 Rotated Factor Correlations Matrix (1969-1990)............................ 154
Table 7.2 Interfactor Correlations (1970-1990)... 166

Preface

There are only a handful of research methodologies that transcend individual fields of enquiry. One such is the subject of this book: Author Citation Analysis (ACA), an application resulting from the combination of multivariate statistical analysis and bibliographic analysis, a field we have come to call bibliometrics. This timely book provides a blueprint for researchers to follow in a wide variety of investigations. The methodology was formerly buried in scholarly articles written mostly by skilled bibliometricians who possess both a mastery of multivariate statistics as well as a full understanding of the significance of bibliographic citations. The generation of the required cocitation matrices was hitherto an expensive venture (because of database and mainframe computer charges), to be undertaken by professional librarians, only some of whom were familiar with the intricacies of the procedure. These matrices can now be generated by users of such bibliographic database portals such as ISI. However, this book goes further by providing an alternative approach which focuses on building custom bibliographic databases and generating corresponding author cocitation matrices. Custom databases allow the researcher to circumvent some of the inherent problems that arise from using commercial databases. Of course, subsequent chapters that explain step-by-step ACA procedures are valid regardless of how the initial cocitation matrix is compiled.

At the heart of ACA lies the notion that authors who build upon the ideas of others will cite them in their articles and are considered to be in the same intellectual sphere. The technique has therefore been used to delineate the intellectual structure of fields, to compare research traditions, to show conceptual differences between competing approaches to software development, and even to demonstrate the existence of a paradigm shift in a particular discipline. It is customary to select a group of at least 30 authors who represent the thinking of the field being investigated. A major hurdle to overcome is the selection of authors to use for the ACA. For well-research fields such as database design, in which there are hundreds of authors, the procedure would probably have to involve the judgments of some of the leading researchers, who could be asked to contribute deserving names to the author list. These names would then have to be individually paired with each one of the other selected authors in the set and run using a suitable commercially available bibliographic database to determine the number of cocitations. Obviously, the task grows exponentially with each additional author. Given the very large number of journals (and therefore articles) in these bibliographic databases, it is a non-trivial search and matching procedure to generate a cocitation matrix for even a modest set of authors. The alternative "objective" approach introduced in this book dispenses with the subjective selection of authors by building much smaller custom databases of journal articles and including each author for further screening. By examining the citation patterns of each article it is possible to produce a much smaller list of authors which is then used to generate the cocitation matrix. Wide availability of the requisite software would certainly facilitate the widespread use of this approach.

Another contribution of this book is the inclusion of a full-blown example of how an ACA was conducted to investigate the intellectual structure of decision support systems (DSS) research. Data from DSS literature over the period 1969-1990 was collected, upon which the ACA was performed. The three standard multivariate analyses (factor analysis, multidimensional scaling and cluster analysis) were applied to the author cocitation matrix derived and seven clusters of DSS research subfields and reference disciplines were discovered. This approach can be extended and fruitfully applied to any number of other fields of knowledge.

It is hoped that this guide to performing ACA will encourage other researchers to explore the intellectual structures of various disciplines in innovative ways, and extract ever more fruit from the vast store of knowledge that has been accumulated over time.

Sumit Sircar
Armstrong Distinguished Professor of Communications Technology & Management
Richard T. Farmer School of Business
Miami University
Oxford, OH 45056

Acknowledgements

Special thanks go to Mr. Praveen Kumar Chevala of ImageScan in Lanham, Maryland, formerly a graduate student in the Computer Science Department of Middle Tennessee State University, for the development of a cocitation matrix generation system. My daughter, Caroline, has done an excellent job in proofreading the manuscript. Chapter 7 is largely based on the two earlier publications from OMEGA and Journal of the American Society for Information Science with Permission from Elsevier Science and Wiley. I am very grateful for the permission to reproduce the articles.

INTRODUCTION

A huge body of knowledge existing today is the result of a cumulative research tradition. Researchers build on each other's and their own previous work. Definitions, topics and concepts are shared and interesting lines of inquiry need to be continuously followed up. To facilitate progress of an academic field, it is important to build such a cumulative research tradition. In this process of knowledge creation, it is necessary to identify, examine, and trace the intellectual linkage to each other in a given academic field as a basis of assessing the current state of its field to guide future development. The intellectual linkages are established through the process of referencing and citation. These intellectual linkages can be systematically examined by means of counting and analyzing the various facets of intellectual activity outputs in the form of written communications.

Over the past 80 years, the way we count and analyze the citation frequency dramatically changed from the early manual transcribing and statistical computation of citation data to computer-based citation data creation and its manipulation. The term statistical bibliography was coined by Hulme in 1922 (Hulme 1923) as a research tool for examining the intellectual development and structure of an academic discipline. Since then, we have seen continuous development of the field of bibliometrics. The principal method of bibliometrics is citation analysis through counting and analyzing the citation frequencies. The most important milestone in the development of citation analysis was established by Garfield. He presented an idea for the management of scientific information using a comprehensive citation index in 1955 and three years later founded the Institute for Scientific Information (ISI)(Garfield 1955). For a detailed description of theory and application of citation indexing, see (Garfield 1979). As of

December 2001, the company has sold 80 different products including *Web of Science, Science Citation Index, etc.* A citation index is a listing of all referenced or cited source items published in a given time span associated with the citing articles.

These citation index files are online bibliographic databases accessible only thorough online-based information service companies such as Dialog®, Profound®, DataStar™, Questel/Orbit, etc. Dialog@Site is a new web-based product that enables the user to search the world's leading databases with web browser. Dialog@site users guide is available at: http://products.dialog.com/products/atsite/pdf/userguide.pdf. Most of the published cocitation analysis is based on the analysis of the first author in the cited articles.

PRIMARY OBJECTIVE OF THIS BOOK

This book introduces an alternative approach to conducting author cocitation analysis (ACA) without relying on commercial citation databases such as index ISI citation index, based on custom bibliographic database and cocitation matrix generation systems specifically developed to use the custom database. The alternative approach can be an effective research tool overcoming several weaknesses of commercial online data-based ACA research.

First, the approach we are introducing here has the capability to access to the non-primary authors of cited references. The non-primary authors refer to all authors other than the first author. The inability to access non-primary authors is a critical shortcoming of ACA research utilizing the commercial databases. Theoretically, the contributions made by non-primary authors must be counted when examining the intellectual structure of an academic discipline.

Second, strict criteria can be applied to the selection of citing articles. A researcher does not always write articles in a specialized field throughout his/her lifetime. Research interests can shift from one subspecialty area to other areas within an academic discipline. Custom bibliographic databases can be built to include only writings in a specific field. Custom database requires hard labor and a time-consuming process to build in the entire process from selecting citing articles, entering cited references from the citing articles.

Third, the alternative approach we introduce here can be a more effective tool for identifying the intellectual structure of an academic field more accurately as well as its reference disciplines. All previous ACA studies except the ones conducted by Eom and his colleagues (Eom 1996a; Eom 1998a; Eom 1998b; Eom 2002; Eom et al. 1996; Eom et al. 1993b) failed to identify the reference disciplines of an academic field. The reason for the failure was the method used to select authors to use for ACA. The method starts a predetermined list of authors selected by subjective judgments of researchers. It is impractical for ACA researchers to include all authors in the reference disciplines of an academic field prior to conducting ACA analysis. If ACA researchers somehow manage to include authors in the reference disciplines of an academic filed, ACA produces empirical maps of prominent authors selected by the researchers. However, with the approach introduced in this book, ACA becomes an exploratory tool. It can dig up the roots (reference disciplines), locate the trunk (foundations of an academic discipline), and sift through branches (subspecialties) of a tree (an academic discipline). The critical element that makes ACA an exploratory tool is the custom bibliographic databases and the author selection method of screening the entire databases to finalize the author set for ACA analysis. This can be called the bottom-up approach. The majority, if not all, of ACA studies using commercial databases are based on the top-down approach – selecting authors applying the subjective judgments prior to ACA analysis. The end result of the top-down approach is simply clustering the subjective author set into several subgroups. With this approach, ACA is inherently a limited tool for identifying the changing structure of an academic field and tracing emerging/fading scholars.

The custom databases can be built to include only writings in a specific domain/subspecialty. Building custom databases requires hard labor and time-consuming process. However, there are important advantages in using custom databases. ISI social science citation index includes bibliographic information, author abstracts, and cited references found in more than 1700 scholarly social science journals covering more than 50 disciplines. Nevertheless, some critically important journals could be missing. For example, in the field of decision support systems, *Decision Support Systems and Electronic Commerce* is a journal dedicated to publish articles in the decision support systems and electronic commerce area. To identify the intellectual structure of the decision support systems area, social science citation index based research could possibly reach inaccurate results due to the exclusion of such an important journal. In this case, building custom databases could be an effective approach.

Fourth, this book describes step-by-step ACA procedures for novice SAS users. This book provides explicit instructions to build bibliographic databases, process them to compile cocitation matrix, prepare SAS input files, and interpret the results. This book provides the reader with a useful, instructional guideline to conduct ACA research regardless of the bibliographic databases used -- in-house databases or commercial citation databases. With commercial citation databases, cocitation matrix can be retrieved without painstaking efforts to create the bibliographic databases. After the retrieval of author cocitation counts, many steps and procedures must still be followed to accomplish the goals of ACA as shown in Figures 3.1 and 4.1. Each and every step is an unstructured process for those inexperienced researchers. This book is intended to help them conduct ACA research.

This book can also be useful for those who are not familiar with three multivariate statistical techniques (factor analysis, cluster analysis, and multidimensional scaling). The book shows elementary procedures to prepare SAS data files, process them, and analyze the outputs. Some of the chore activities must be learned from trial and error, which is often time-consuming and frustrating. I have learned much from my mistakes and experience. Even to those who are not ACA researchers the book provides useful tips on each process of research using multivariate techniques. Although I have included basics of SAS programs for three multivariate statistical analysis techniques (factor analysis, cluster analysis, and multidimensional scaling), this introduction is not intended to give a comprehensive one-step guideline for ACA students. It is an introduction of multivariate statistical techniques using the SAS system to analyze cocited author counts. With this introduction, ACA students are in a better position to study SAS language and procedures, SAS graph software, and SAS/STAT users' guide. The sample SAS programs in the book are working programs that can be used with different data sets.

ORGANIZATION OF THE BOOK

The book consists of seven chapters. Chapter 1 discusses basic concepts of bibliometrics, author cocitation analysis along with ACA's assumptions, purposes, benefits, limitations, and criticisms. Chapter 2 discusses the first step in ACA, building bibliographic databases to include citing (source) and cited references. Chapter 3 overviews ACA steps. They include selection of authors, retrieval/generation of paired author cocitation frequencies, preparing inputs to the SAS system, multivariate statistical analyses of author cocitation matrix, and validations/interpretation of SAS outputs. The input to the SAS system is cocitation frequency matrix. The data matrix is processed by three multivariate procedures (factor, cluster, and multidimensional scaling). Chapters 4 and 5 deal with factor analysis and cluster analysis respectively. In conducting cluster analysis, two macro statements are necessary to convert the frequency matrix to distance matrix since a higher frequency among authors must be translated into a closer proximity.

Chapter 6 focuses on the multidimensional scaling (MDS) which produces a coordinate matrix, not graphical outputs. Therefore, additional PLOT and G3D procedures are necessary to produce two dimensional plots and annotated 3D scatter plots. Throughout this book, we use a small data set of 9 variables. Advantages of using such small number of variables include a clearer understanding of data preparation step and an easier interpretation of outputs. On the other hand, smaller data set may make it difficult to fully demonstrate problems that can arise with a large number of variables such as scree plot, finding optimal number of factors based on the factor interpretation, etc. Chapter 7 introduces an ACA study in the management information systems area to demonstrate some concepts that cannot be adequately explained with the smaller dataset used in chapters 4, 5, and 6.

CHAPTER 1 BASIC CONCEPTS OF BIBLIOMETRICS AND AUTHOR COCITATION ANALYSIS

Author cocitation analysis (ACA) is a research tool whose idea originated in the late 1960s (Rosengren 1968). A series of papers from researchers at the College of Information Studies at Drexel University have made ACA a popular research tool in the area of library science (White 1981; White 1983; White et al. 1981; White et al. 1982). This chapter provides basic concepts of bibliometrical research tools in order to understand an alternative approach we are introducing.

WHAT IS BIBLIOMETRICS (STATISTICAL BIBLIOGRAPHY)?

The term statistical bibliography was coined by E. Wyndham Hulme in 1922 (Hulme 1923). The purposes of statistical bibliography are described as:

> 1. to shed light on the processes of written communication and of the nature and course of development of a discipline (in so far as this is displayed through written communication), by means of counting and analyzing the various facets of written communications (Prichard 1969).
> 2. the assembling and interpretation of statistics relating to books and periodicals ... to demonstrate historical movements, to determine the national or universal research use of books and journals, and to ascertain in many local situations the general use

of books and journals (Raisig 1962).

Pritchard (1969) suggested using the term bibliometrics instead of statistical bibliography. He believed that the term statistical bibliography was vague and could be confused with statistics itself or bibliographies on statistics. According to Pritchard, bibliometrics is defined as "the application of mathematics and statistical methods to books and other media of communication." Citation analysis is a major type of bibliometrics (statistical bibliography).

CITATION ANALYSIS

Knowledge creation and dissemination in a discipline are facilitated through the circulation of ideas among "invisible colleges" (Crane 1972). Each individual contributes to the body of knowledge by building on what others have already accomplished. In this process, referencing and citation are important tools to link each other's writing.

Citation analysis can be basically classified into two basic types. The first type is counting citation of a document or set of documents authored by an individual without considering intellectual linkage. The second is co-citation analysis of authors or documents to identify intellectual linkages among authors/publications. For examples of the first type of analysis, see (Eom 1994; Eom et al. 1993a). Such citation analysis is often used to compare research productivity of individual faculty member and/or university specific academic program measured by citation counts.

The next type is the cocitation analysis of multiple authors or multiple documents, which was developed under the name of "co-mentions analysis" in 1968 (Rosengren 1968; Rosengren 1990). Systematic analysis of co-citation can be done using many different methods including bibliographic coupling, document co-citation analysis, author co-citation analysis, and co-word analysis. Since the primary purpose of the book focuses on author cocitation analysis, we will focus on only one of many available tools here. For a methodological review of these four different methods, see Baker (1990).

AUTHOR COCITATION ANALYSIS

There are two primary types of cocitation analysis to map the intellectual structure of an academic field: document cocitation analysis and author cocitation analysis (ACA). Document cocitation analysis involves the analysis of a set of selected documents (e.g., journal articles, books, proceedings, etc.) in terms of which pairs of documents are cited together. Author cocitation analysis, introduced in 1981, is a more general approach to identify, examine, and trace the intellectual structure of an academic discipline by counting the frequency with which any work of an author is cited to any work by another author in the references of citing documents (Bayer et al. 1990).

ACA, a major area of bibliometrics, is a technique of bibliometrics that applies quantitative methods to various media such as books, journals, conference proceedings, and so on. ACA is "a set of data gathering, analytical, and graphical display techniques that can be used to produce empirical maps of prominent authors in various areas of scholarship" (McCain 1990).

The cocitation of authors occurs when a citing paper cites any work of authors in reference lists. Many information scientists and author cocitation analysis researchers define an author as "a body of writings by a person" or "a body of contributions by a person." The term "contributions" may be better since it can include any type of contribution that can be cited as a reference such as speeches delivered at professional meetings, personal communications including conversation and letters, and other media. The term "a person" refers to a single

Figure 1.1 Can You See the Trunk, Branches, and Roots of the Tree?

Figure 1.2 ACA: A Tool for Digging Up the Roots, Examining the Trunk, and Identifying the Branches

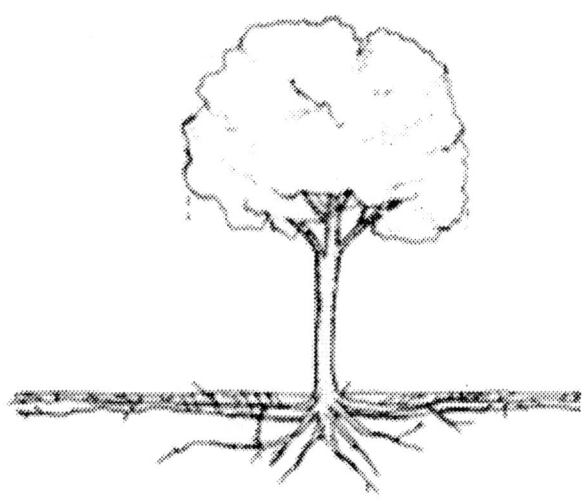

author or one of multiple authors. These different uses of terms are related to citation databases used in the study. Most commercial citation databases and software access only the first author, regardless of the number of multiple authors, when retrieving author cocitation counts. This has been the critical weakness of using the commercial citation databases and software. However, this book is based on the bibliographic database I have created, which includes all contributions such as speeches delivered at various meetings and software we have developed that can access all multiple authors. With custom-built bibliographic databases, and the bottom-up approach of the selection of author sets, ACA becomes an exploratory tool for digging up the roots (reference disciplines), locating the trunk (foundations), and sifting through the branches (subspecialties) of a tree (an academic discipline). The critical element that makes ACA an exploratory tool is the custom bibliographic databases and the author selection method of screening the entire databases to finalize the author set for ACA analysis. For an overview and discussion of the continuing relevance of ACA to the study of the intellectual structure of literatures, see a special issue of Journal of the American Society for Information Science, vol. 41, no. 6, 1990. The issue contains a brief introduction by Howard D. White (Guest Editor) and a technical overview of the steps in ACA (McCain 1990).

ASSUMPTIONS OF AUTHOR COCITATION ANALYSIS

Author cocitation analysis is based on the assumptions that "bibliographic citations are an acceptable surrogate for the actual influence of various information sources" (McCain 1986) and that the cocitation analysis of a field yields a valid representation of the intellectual structure of the field (Bellardo 1980; McCain 1984; McCain 1990; Smith 1981). According to Bellardo (1980), the fundamental premise of cocitation analysis is that the greater the frequency a pair of documents/authors are cited together, the more likely it is that they are related in content. The cocitation frequency of authors represents relationships between authors. Authors whose works are cited together frequently are interpreted as having close relationships with one another. ACA is based on the assumptions that "cocitation is a measure of the perceived similarity, conceptual linkage, or cognitive relationship between two cocited items (documents or authors)" and "cocitation studies of specialties and fields yield valid representations of intellectual structure" (McCain 1986).

PURPOSES AND BENEFITS OF AUTHOR COCITATION ANALYSIS

Citation analysis is often used to determine the most influential scholars, publications, or universities in a particular discipline by counting the frequency of citations received by *individual* units of analysis (authors, publications, etc.) over a period of time from a particular set of citing documents. However, citation analysis cannot establish relationships among units of analysis. ACA is the principal bibliometric tool to establish *relationships* among authors in an academic field and thus identify subspecialties of a field and how closely each subgroup is related to each of the other subgroups. By establishing relationships among authors, ACA provides a basis of revealing the intellectual structure of literature and defining the principal subject (major area of subspecialties in an academic discipline and their contributing disciplines) through the empirical consensus of numerous authors in an academic discipline.

In her landmark ACA based research which examined the intellectual evolution and development of the MIS area, Culnan (1986, p.156) discusses the importance of the study of the intellectual development of a field of study:

> Researchers in any academic discipline tend to cluster into informal networks, or "invisible colleges," which focus on common problems in common ways (Price 1963). Within these networks, one researcher's concepts and findings are soon picked up by another to be extended, tested and refined, and in this way, each person's work builds on that of another. The history of

exchanges between members of these subgroups in a discipline describes the intellectual history of the field. .....
Researchers can benefit by understanding this process and its outcomes because it reveals the vitality and the evolution of thought in a discipline and because it gives a sense of its future. In a relatively new field such as MIS, this understanding is even more beneficial because it identifies the basic commitments that will serve as the foundations of the field as it matures....

LIMITATIONS /CRITICISMS OF AUTHOR COCITATION ANALYSIS

ACA is a quantitative tool that cannot and should not be used by itself to determine the intellectual structure of academic disciplines. This is a supporting quantitative tool that must be used with further qualitative analysis of bibliographic data.

While citation analysis can be a useful research tool due to unobtrusive, precise, and objective characteristics, there are limitations of ACA stemming from the citation behavior of authors and bibliographic databases. In regard to citation behavior of authors, Smith (1981, p. 84) enumerated fifteen reasons for citation based on the work of Garfield (1965).

1. Paying homage to pioneers
2. Giving credit for related works (homage to peers)
3. Identifying methodology, equipment, etc.
4. Providing background reading
5. Correcting one's own work
6. Correcting the work of others
7. Criticizing previous work
8. Substantiating claims
9. Alerting to forthcoming work
10. Providing leads to poorly disseminated, poorly indexed, or uncited work
11. Authenticating data and classes of fact-- physical constants, etc.
12. Identifying original publications in which an idea or concept was discussed
13. Identifying original publications or other works describing an eponymic concept or term..
14. Disclaiming work or ideas of others (negative claims)
15. Disputing priority claims of others (negative homage).

Many problems can also arise in relation to the sources of citation data and mechanics of deriving citations from existing citation indexes. The problems may stem from multiple authorship, self-citations, homographs, synonyms, the unification problems, etc. (Lindsey 1980; Long 1980; Smith 1981). Use of SSI and SCI can raise a potential problem since these sources can exhibit English language bias (Baker 1990). Use of custom databases and the cocitation matrix generation system we developed can eliminate many of the problems discussed above such as multiple authorship, homographs, synonyms, etc.

CHAPTER 2 BUILDING THE BIBLIOGRAPHIC DATABASES

The cocitation matrix generation system accepts two files as major inputs-- citing articles database and cited reference databases. Citing sources and the majority of cited references are from journals. Cited references come from books, conference proceedings, papers presented at the meetings without proceedings, doctoral dissertations and masters thesis, working papers, articles from encyclopedia, book chapters in edited books, newspaper articles, reports, personal communications, electronic sources primarily published on the Internet, unpublished working papers, etc. If hard disk storage requirements are not a major consideration, the database can include all unique fields from each source type. For example, journal article may consist of author (last_name, first_name, middle_name for primary and secondary authors), title of article, journal, Volume, issue, year, date, etc. Book entry may require different fields such as author (last_name, first_name, middle_name for primary and secondary authors), title of books, publisher, city, edition, year, ISBN, etc. Combining all unique fields from each of different reference sources may require more than 100 fields in the database. Most of fields, however, will be "empty". Therefore, there are tradeoffs between storage requirement considerations and maintenance of accurate field names.

DATABASE DESIGN

Considering storage requirements as the primary factor, the bibliographic database we have created consists of the following structures. It is possible to design multiple tables to create a view (interrelated tables). However, here we introduce how to build single files for citing articles and cited references without the data normalization process.

THE STRUCTURE OF THE BIBLIOGRAPHIC FILE

Table 2.1 The Structure of Cited Reference File

Field	Field Name	Type	Width	Dec
1	Lastname	Character	18	
2	Firstname	Character	16	
3	Middlename	Character	100	
4	Title	Character	140	
5	Journal	Character	100	
6	Volume	Character	6	
7	Number	Character	5	
8	Date	Character	13	
9	Year	Character	4	
10	Pages	Character	15	
11	IDNO	Character	4	
12	Frequency	Numeric	3	
13	Citingby	Character	180	
14	Control	Character	3	

Since all citing articles and the majority of cited reference articles are from journals, the database structure is based on the structure suited for the journals articles. When entering bibliographical information other than journals, the "journal" field is used to enter publishers, city, state, etc. Typical bibliographical information is entered in fields 1 through 10. The remaining fields are important for ACA purposes.

IDNO

This field represents citing article's identification numbers. Each citing article is assigned a unique number which can be numeric or alphanumeric. The example shown below includes F8 as idno of the citing article.

FREQUENCY

The frequency field is a three digit numeric field to record the citing frequencies of cited references. The example below also shows that this article entitled "A Survey of Decision Support System Applications (1971-April 1988)" is cited 23 times.

CITINGBY

This is the field that keeps a listing of citing articles' reference numbers. The example record indicates that the article is cited by 23 citing articles with idno from 361 through h6.

Figure 2.1 An Example of Cited Reference Record

```
Lastname    Eom
Firstname   Hyun
Middlename  B. and Sang M. LeeSM
Title       A Survey of Decision Support System Applications (1971-April 1988)
Journal     Interfaces
Vol         20
No          3
Year        1990
Months      May-June
Pp          65-79
Copyok      T
Dssfreq     23
Dssidno     F8
Esidno
Keyword
Dscitingby  361 401 417 438 489 793 837 944 859 902 903 937 939 999 a47 b3 b61 b65 b86 c78 Q32 F31 h6
Esfreq      0
Escitingby
Control     mp
Refno       22
Iloan       F
```

POPULATING DATABASE

SELECTION CRITERIA FOR CREATING THE CITING AND CITED ARTICLE DATABASE

One advantage of the custom-databased ACA is the inclusion of articles selected by researchers. Article and journal selection can possibly affect the outcome of the ACA analysis. The ISI® *Social Sciences Citation Index* (*SSCI*®) and *Social SciSearch*® provide access to current and retrospective bibliographic information and cited references found in over 1,700 of the world's leading scholarly social sciences journals covering more than 50 disciplines. Nevertheless, it is possible that the ISI database may not include all important journals and articles in a discipline. To build a valuable bibliographic database, it is important to set selection criteria when deciding which papers to include in the database. For example, see the data and research methodology section of chapter 6 of the book.

REFERENCED (CITED) ARTICLE DATABASE

As an example of ACA databases, we describe a bibliographic database built for our continuing research. A total of 1,616 *citing* articles contain 25,339 *cited* unique reference records. Each cited record has fields (dssfreq, dsscitingby) to include a total number of frequencies of citation by the citing articles and the list of identification numbers of citing articles in the decision support systems area. Each of the 1616 citing records has an average of 30 references, computed by total citation frequency/total citing records (48556/1616).

Table 2.1 shows that about 75% of cited articles were cited by the citing articles just once. Articles with two citations constitute approximately 13% of total cited records (25,339). Only less than 2% of cited articles (502 articles) have been cited 10 or more times by the 1616 citing articles. Although there are no comparable statistics in other disciplines, the majority of cited articles (about 88% of total cited articles) were cited just once or twice by the 1616 citing articles.

This fact can be a possible symptom of a fragmented research trend in the DSS area, or it could be a quality level indicator of published DSS and its reference discipline research in general.

Table 2.2 Citation Behavior

Citing Freq.	No. of Records	Proportion	Cumulative	Total Freq.	Proportion	Cumulative
1	19077	75.29%	75.29%	19077	39.29%	39.29%
2	3095	12.21%	87.50%	6190	12.75%	52.04%
3	1147	4.53%	92.03%	3441	7.09%	59.12%
4	560	2.21%	94.24%	2240	4.61%	63.74%
5	348	1.37%	95.61%	1740	3.58%	67.32%
6	212	0.84%	96.45%	1272	2.62%	69.94%
7	178	0.70%	97.15%	1246	2.57%	72.51%
8	124	0.49%	97.64%	992	2.04%	74.55%
9	96	0.38%	98.02%	864	1.78%	76.33%
10	71	0.28%	98.30%	710	1.46%	77.79%
11	48	0.19%	98.49%	528	1.09%	78.88%
12	29	0.11%	98.60%	348	0.72%	79.59%
13	43	0.17%	98.77%	559	1.15%	80.75%
14	28	0.11%	98.88%	392	0.81%	81.55%
15	29	0.11%	99.00%	435	0.90%	82.45%
16	21	0.08%	99.08%	336	0.69%	83.14%
17	18	0.07%	99.15%	306	0.63%	83.77%
18	13	0.05%	99.20%	234	0.48%	84.25%
19	14	0.06%	99.26%	266	0.55%	84.80%
20	18	0.07%	99.33%	360	0.74%	85.54%
21	12	0.05%	99.38%	252	0.52%	86.06%
22	9	0.04%	99.41%	198	0.41%	86.47%
23	13	0.05%	99.46%	299	0.62%	87.09%
24	10	0.04%	99.50%	240	0.49%	87.58%
25	5	0.02%	99.52%	125	0.26%	87.84%
26	6	0.02%	99.55%	156	0.32%	88.16%
27	9	0.04%	99.58%	243	0.50%	88.66%
28	6	0.02%	99.61%	168	0.35%	89.00%
29	8	0.03%	99.64%	232	0.48%	89.48%
30	4	0.02%	99.65%	120	0.25%	89.73%
31	6	0.02%	99.68%	186	0.38%	90.11%
32	6	0.02%	99.70%	192	0.40%	90.51%
33	10	0.04%	99.74%	330	0.68%	91.19%
34	2	0.01%	99.75%	68	0.14%	91.33%
35	0	0.00%	99.75%	0	0.00%	91.33%
36	2	0.01%	99.76%	72	0.15%	91.48%
37	3	0.01%	99.77%	111	0.23%	91.70%
38	3	0.01%	99.78%	114	0.23%	91.94%
39	5	0.02%	99.80%	195	0.40%	92.34%
40	2	0.01%	99.81%	80	0.16%	92.51%
41	2	0.01%	99.81%	82	0.17%	92.67%
42	1	0.00%	99.82%	42	0.09%	92.76%
43	2	0.01%	99.83%	86	0.18%	92.94%
44	1	0.00%	99.83%	44	0.09%	93.03%
45	3	0.01%	99.84%	135	0.28%	93.31%
46	2	0.01%	99.85%	92	0.19%	93.50%
47	1	0.00%	99.85%	47	0.10%	93.59%

Continued on next page

Table 2.2---*Continued*

Citing Freq.	No. of Records	Proportion	Cumulative	Total Freq.	Proportion	Cumulative
48	2	0.01%	99.86%	96	0.20%	93.79%
49	4	0.02%	99.88%	196	0.40%	94.19%
50	2	0.01%	99.89%	100	0.21%	94.40%
51	2	0.01%	99.89%	102	0.21%	94.61%
52	1	0.00%	99.90%	52	0.11%	94.72%
53	0	0.00%	99.90%	0	0.00%	94.72%
54	0	0.00%	99.90%	0	0.00%	94.72%
55	0	0.00%	99.90%	0	0.00%	94.72%
56	2	0.01%	99.91%	112	0.23%	94.95%
57	2	0.01%	99.91%	114	0.23%	95.18%
58	0	0.00%	99.91%	0	0.00%	95.18%
59	0	0.00%	99.91%	0	0.00%	95.18%
60	1	0.00%	99.92%	60	0.12%	95.31%
61	0	0.00%	99.92%	0	0.00%	95.31%
62	1	0.00%	99.92%	62	0.13%	95.43%
63	0	0.00%	99.92%	0	0.00%	95.43%
64	1	0.00%	99.93%	64	0.13%	95.57%
65	0	0.00%	99.93%	0	0.00%	95.57%
66	1	0.00%	99.93%	66	0.14%	95.70%
67	0	0.00%	99.93%	0	0.00%	95.70%
68	0	0.00%	99.93%	0	0.00%	95.70%
69	1	0.00%	99.93%	69	0.14%	95.84%
72	2	0.01%	99.94%	144	0.30%	96.14%
75	1	0.00%	99.94%	75	0.15%	96.29%
77	1	0.00%	99.95%	77	0.16%	96.45%
85	1	0.00%	99.95%	85	0.18%	96.63%
88	1	0.00%	99.96%	88	0.18%	96.81%
97	2	0.01%	99.96%	194	0.40%	97.21%
103	2	0.01%	99.97%	206	0.42%	97.63%
104	1	0.00%	99.98%	104	0.21%	97.85%
116	1	0.00%	99.98%	116	0.24%	98.09%
118	1	0.00%	99.98%	118	0.24%	98.33%
144	1	0.00%	99.99%	144	0.30%	98.63%
189	1	0.00%	99.99%	189	0.39%	99.02%
236	1	0.00%	100.00%	236	0.49%	99.50%
242	1	0.00%	100.00%	242	0.50%	100.00%

Cited Record   25339
Total Citation 48556   Average Citing Frequencies 30.04703
Citing Records 1616

Figure 2.2 Citation Pattern

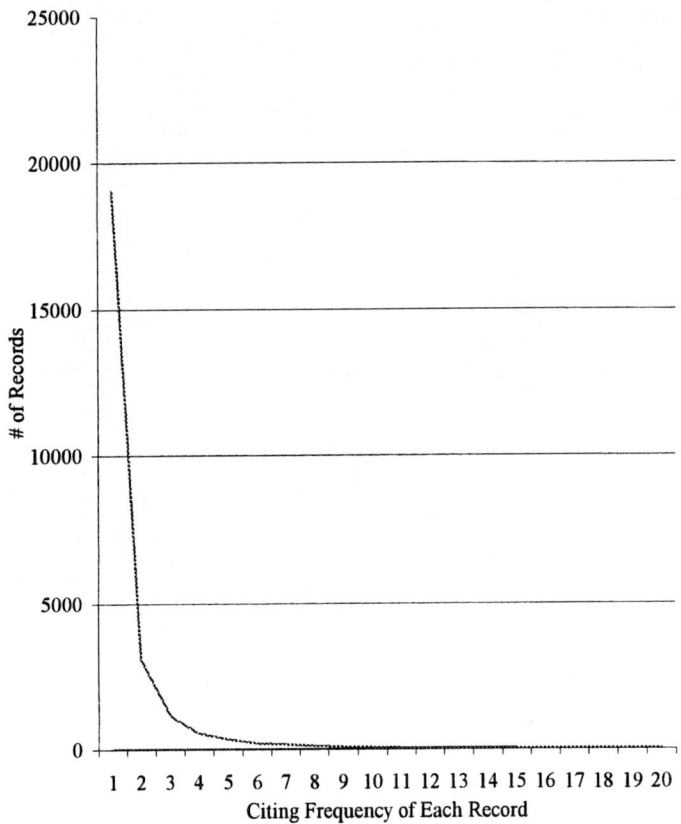

CHAPTER 3 OVERVIEW OF AUTHOR COCITATION ANALYSIS STEP

ACA consists of the assembling and interpretation of bibliographical statistics taken from the cited references which are taken from the selected citing articles. (See, Figure 3.1).

1. Selection of Author
2. Collection of Cocitation Frequencies
3. Preparation of SAS Input Files
4. Multivariate Statistical Analysis of Correlation Matrix
5. Output Preparation
6. Validation and Interpretation

SELECTION OF AUTHORS/COCITATION THRESHOLDS

The goal of ACA is to produce empirical maps of prominent authors in an academic discipline. In doing so, the first step is to select eminent scholars in the academic discipline. Basically, two approaches are available: starting a predetermined list of authors in a given field (the subjective approach), or compiling a set of authors from scratch (the objective approach). An important purpose of ACA is an overall examination of the intellectual structure of an academic discipline. Therefore, it is critical to establish a diversified list of authors. McCain (1990, p.433) states:

In the aggregate, this author set *defines* the scholarly landscape being mapped. If the authors are not chosen to capture the full range of variability in subject specializations, methodologies, political orientations, etc., these aspects of structure cannot be determined.

SUBJECTIVE (TOP-DOWN) APPROACH

The first approach may start with a pre-determined list of authors or with the selection of a list of authors to be searched as cited references in the ISI databases as discussed by McCain (1990). The predetermined list of authors can be compiled by:

- Personal knowledge
- Consultation with researchers in the area to be studied
- Conducting surveys
- Using directories
- Organizational membership
- Conference attendance rosters
- Lists of awards, etc.

Compiling a predetermined list of authors inevitably involves subjective judgments. This approach can be efficient in that no lengthy time is spent to finalize a list of authors for further analysis. A critical weakness of this approach is that it may often fail to identify emerging scholars in a given area of an academic discipline. The majority of previous research in this area has used the subjective approach of using a predetermined list of eminent scholars (Culnan 1986; McCain 1986). The reason for doing so was not because this approach is superior but because most of the previous research used commercial online bibliographic databases to retrieve the cocitation frequency matrix.

Figure 3.1 Author Cocitation Analysis Steps

1. Selection of Authors
   Subjective Approach (1.1 -- 1.3)
   Objective Approcah
   1.4 A premininary list of authors through loose screening
   1.5 Finalizing the author set

2. Generation/Compilation Cocitation Frequencies among the Authors selected from Step 1.

3. Input Data (Cocitation Matrix) Preparation
   3.1 Initial Cocitation Matrix with Adjusted Diagonal Cells
   3.2 Transposed Cocitation Matrix
   3.3 SAS Input File
       3.3.1 Factor Analysis
       3.3.2 Cluster Analysis
       3.3.3 MDS 2-D Analysis
       3.3.4 MDS 3-D Analysis

4. Multivariate Analysis of Correlation Matrix
   4.1 Factor Analysis
       4.1.1 Factor Structure Analysis
       4.1.2 Factor Correlations Analysis
   4.2 Cluster Analysis
   4.3 Multidimensional Scaling
       4.3.1 Two-Dimensional MDS
       4.3.2 Three-Dimensional MDS

5. Output Preparation
   5.1 Rotated Factor Pattern Tables
   5.2 Factor Structure Correlation Tables
   5.3 Inter-factor Correlation Tables
   5.4 Cluster Analysis Output (Dendrogram)
   5.5 Two-Dimensional MDS Graph
   5.6 Three-Dimensional MDS Graph

6. Validation and Interpretation
   6.1 Statistical Validation
   6.2 Consultation with Field - Specialists
   6.3 Interpretation based on Researchers' Judgment

OBJECTIVE (BOTTOM-UP) APPROACH

The objective approach starts with no predefined list of authors at all. No subjective judgments are necessary. It will be very difficult to apply this objective approach to the commercial databases such as social science citation index due to the size of the databases and high costs. However, it is relatively manageable to screen the whole custom-built databases to select a list of authors for further analysis. ACA, using custom databases, allows researchers to unobtrusively select authors for the study. Emerging scholars are more likely to be included in the selection process, unlike the subjective approach discussed earlier.

The other important advantage of using custom-built databases includes the identification of reference disciplines. It will be very difficult to include all predefined list of authors in the reference disciplines prior to conduct actual analysis, if not impossible due to the search time and financial constraints. As explained below, the objective approach we are introducing is suitable when using the custom-built databases, since researchers do not have to start with a list of authors in the reference disciplines.

THE LOOSE SCREENING TO PRODUCE A PRELIMINARY LIST

Unlike ACA using commercial on-line databases such as ISI citation index files, the selection process starts with no predefined list of authors. Instead, each author in the in-house bibliographic databases is initially included for further screening. The first stage of loose screening is based on the citation frequency of each record. Table 2.2 reveals an interesting citation pattern.

* About 75% of all cited works are cited just once.
* Only about 7% of all cited works are cited three or more times.
* About 5% of all cited works are cited five or more times.

Based on this citation pattern of the database, deleting all records whose citation frequencies are ten or less (98.31%) we have now less than 2% of all cited reference records for further analysis. Therefore, this stage yields a list of a significantly reduced number of authors. The actual preliminary list of authors can be compiled via identifying unique names in smaller number of records. These records may contain different numbers of unique authors, whether primary or non-primary, after removing redundant names from the first author fields of each record and adding names from the non-first author fields (the middle name fields in our database).

<u>FINALIZING THE LIST OF AUTHOR VIA DETAILED ANALYSIS</u>

Citation counts of the individual authors in the preliminary list can be used to finalize the list of authors. This step further filters the preliminary list compiled from the previous step into a set of authors based on author *cocitation* counts (frequencies). Although it rarely happens, theoretically it is possible for any authors with higher citation counts to have very low cocitation counts with other authors. In this case, these authors may appear in the final author set. But they will carry lower factor loadings that can be interpreted as insignificant authors to the formation of the intellectual structure of the academic discipline under study. To accurately and objectively examine the intellectual structure of a discipline, personal judgment must be avoided in selecting authors by objectively counting the frequency of each name from the data bases.

Using in-house databases, author selection criteria must be established in regard to the number of citation received. As the citation behavior figure shows, the higher number of threshold means the smaller number of authors for further analysis. The optimal number of authors is primarily dependent on the number of cited database records. In addition, there may be some differences in the citation behavior across the different academic disciplines. There are no quantitative tools that can be blindly applied in deciding the number of authors. A study of author cocitation of a journal in consumer research over the 15 year period used 4 citation as selection criteria to compile the list of authors (Hoffman et al. 1993). Other studies of ACA to map the intellectual structures of decision support systems, the adjusted diagonal cell values of 25 or more were used (Eom 1995; Eom 1996a). Once the threshold values are used to select the author list, it is important to apply the same criteria consistently to the subsequent studies to be followed to trace the changing structures of a discipline.

AD HOC CRITERIA FOR FURTHER SCREENING OF AUTHORS

Due to the possible instability of small cocitation counts, author cocitation analysis researchers introduced several ad hoc criteria for further screening a large pool of candidate authors to finalize a list of authors. The criteria include a *mean cocitation rate* above a certain lower limit per author in each time period (e.g., nine for 10 years of Social Scisearch data), cocitation with at least one-third of the entire author set, or restricting the final author set to the 20% receiving the highest number of citations and cocitations in initial retrieval trials. For further details on several different approaches to compiling a predetermined list of authors, see (McCain 1990). However, author cocitation analysis researchers suggested that all these criteria be applied to the commercial on-line databases such as SCISEARCH and SOCIAL SCISEARCH. The nature of the database we created makes it meaningless to use those criteria suggested by ACA researchers.

Our databases are significantly different from those commercial databases in terms of record size. Besides, the cocitation matrix generation system we developed gives access to cited coauthors as well as first authors. Due to these differences, we could not follow the suggested criteria by McCain (1990), such as "a mean cocitation rate of 'x' or more cocitations in each time period." Rather, we had to invent a new criterion through the method of trial and error. We experimented with the sensitivity of changing the cocitation threshold on the final outcomes (number of meaningful factors to accurately represent research subspecialties). With our databases, we conclude that the number of cocitations of an author with himself/herself can be a better criterion to determine the final author set due to its simplicity. Applying the mean cocitation criteria to worksheet files of any spreadsheet programs (the output from the cocitation counts generation system) involves too many computations. For example, whenever we delete/add an author to the final author set, we need to compute the mean cocitation rate of each author again. Using the cocitation rate of 25 with himself/herself in the investigation period, the final set of authors was chosen. In our previous papers (Eom 1995; Eom et al. 1996), the cocitation rate of 25 with himself/herself was applied to finalize a list of authors for further analysis.

Regardless of the nature of bibliographic databases (commercial vs. custom-built), determining the threshold cocitation rate is not the result of a structured process; rather, it is an unstructured process requiring the investigator's personal judgments. An exact quantitative basis for deciding the threshold cocitation rate has not been developed. Lowering the threshold in general increases the number of authors to be included in a study, which in turn may or may not change the number of meaningful factors in the study. Also, it is important to point out that cocitation thresholds themselves, as sole connection criteria, are suspect in a highly multidisciplinary area. One should look at the overall connectedness and the focused cocitation counts as well.

GENERATION OF COCITED AUTHOR COUNTS

Cocitation counts can be either retrieved from commercial online bibliographical databases such as *Science Citation Index, Social Sciences Citation Index*, and *Arts and Humanities Citation Indexes,* or generated from the custom-built bibliographical databases. If researchers use the commercial on-line databases, cocitation frequencies will be retrieved using the query command. Several examples of those commands are given in McCain (1990).

THE COCITATION COUNT GENERATION SYSTEM

This book introduces an alternative approach to ACA research -- the generation of cocited author counts using a custom-built bibliographic database and in-house cocitation count generation systems. FoxBase database management systems are used to enter the bibliographic records. The system is coded using Fox-Base database management systems. It computes author cocitation frequencies between any pair of all (primary and non-primary) authors under study. The author cocitation frequency generation system enables the users to overcome the problem with the Institute for Scientific Information (ISI) databases which code only the first author of a cited work. The cocitation matrix generation system we developed gives access to cited coauthors as well as first authors.

Figure 3.2 Main Menu of Author Cocitation Frequency Generation Systems

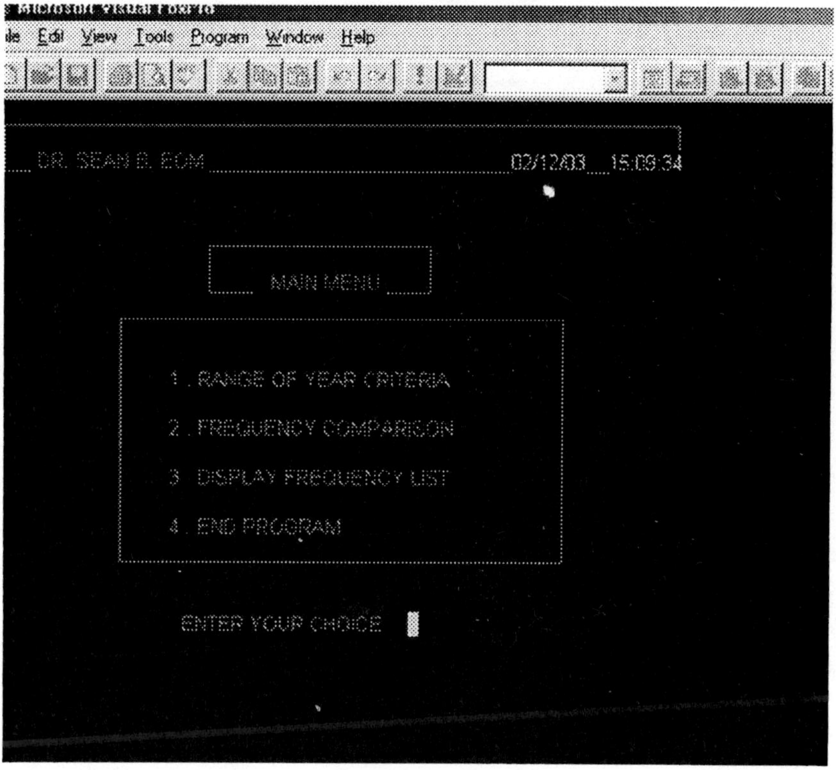

Figure 3.3 Retrieval of Cocitation Frequency between Two Authors

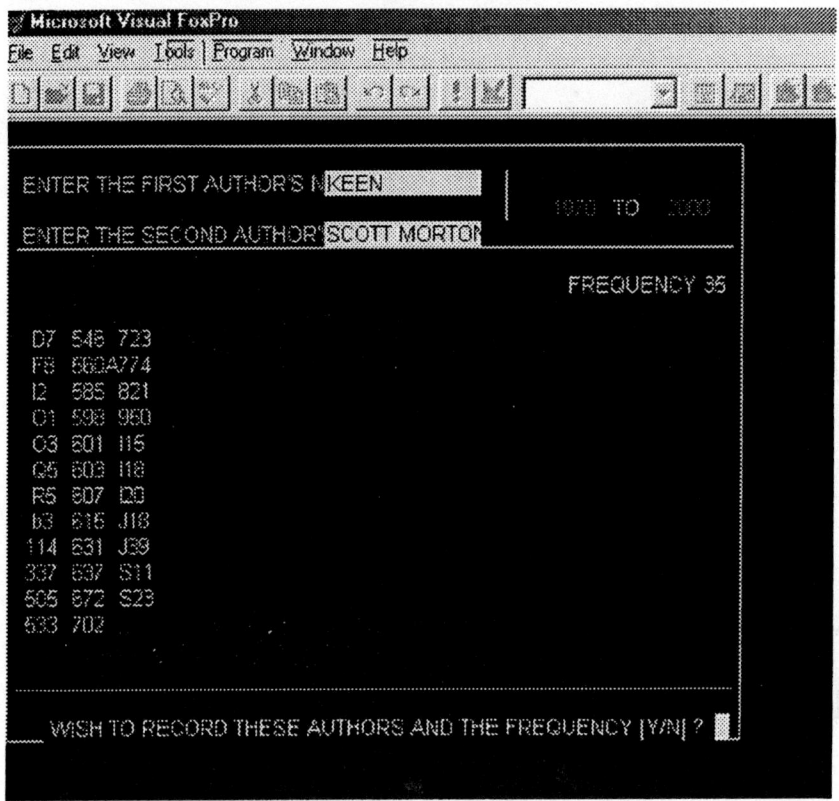

Figure 3.4 Displaying Cocitation Frequency of a File

AUTHOR1	AUTHOR2	FREQUENCY
ACKOFF	GREENBERG	2
ACKOFF	GROHOWSKI	7
ACKOFF	GUSTAFSON	1
ACKOFF	HACKMAN	7
ACKOFF	HEMINGER	9
ACKOFF	HENDERSONJ	7
ACKOFF	HILTZ	9
ACKOFF	HOGARTH	5
ACKOFF	HOLLINGSHEAD	1
ACKOFF	HOLMES	1
ACKOFF	HOLSAPPLE	12
ACKOFF	HUBER	17
ACKOFF	HURT	2
ACKOFF	IVES	4
ACKOFF	JANIS	4
ACKOFF	JARKE	9
ACKOFF	JARVENPAA	5
ACKOFF	JELASSI	46

PRESS ANY KEY TO VIEW THE NEXT SCREEN ....

Figure 3.5 Cocitation Frequency Matrix Preparation Steps in ACA

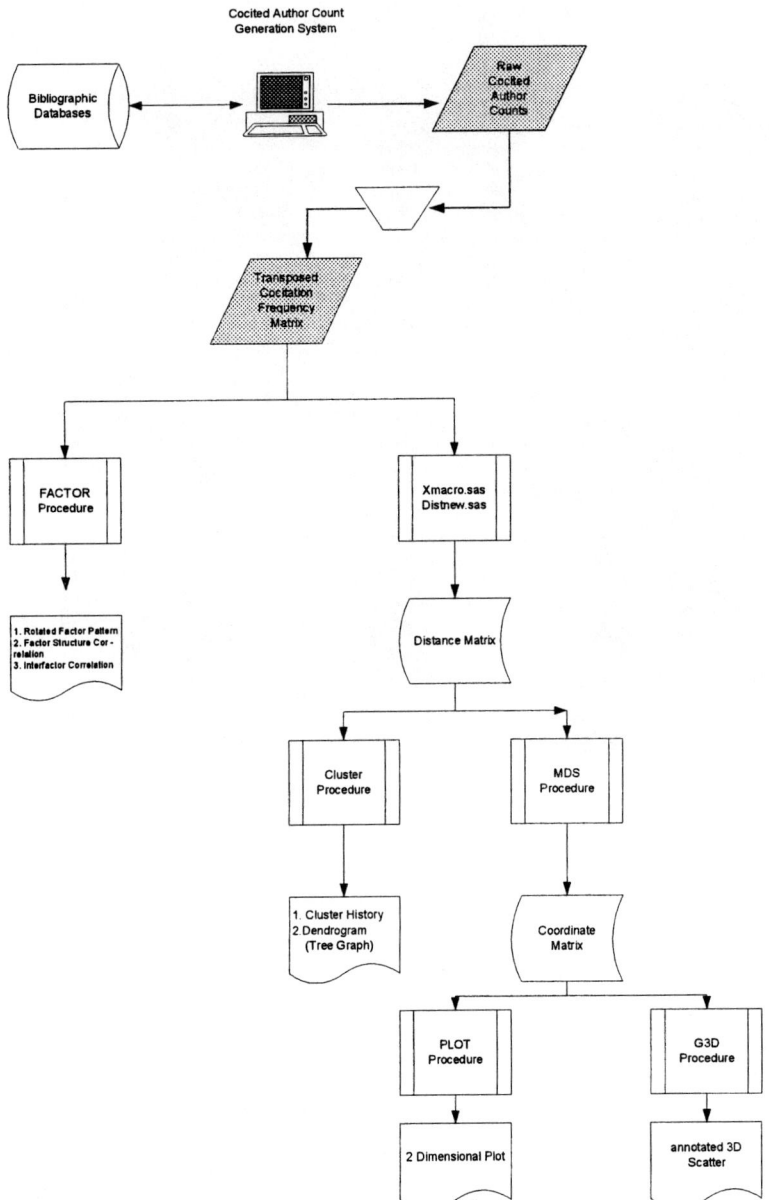

RAW COCITED AUTHOR COUNTS

In ACA, cocitation frequency (cocited author counts) is the prime input data. The cocitation count generation system produces a raw author cocitation frequency table such as Table 3.1. If the number of authors under study is n, all possible number of cocitation counts to be produced by the system is computed by n*(n+1)/2. For example, Table 3.2 lists 45 cocitation counts with 9 authors. The 45 cocitation counts include 9 diagonal cell values. On the other hand, the possible number of cocitation counts to be retrieved by the commercial system is computed by n*(n-1)/2. There is a difference between the two because the system we developed produces diagonal cell values while the diagonal values are not retrieved by the commercial system.

Table 3.1 Part of The Cocitation Count Generation System Output

AUTHOR1	AUTHOR2	COCITATION FREQUENCY
ALTER	ALTER	121
ALTER	BONCZEK	25
ALTER	CARLSON	66
ALTER	HUBER	23
ALTER	KEEN	92
ALTER	SCOTTMORTON	83
ALTER	SIMON	38
ALTER	SPRAGUE	67
ALTER	WHINSTON	27
BONCZEK	BONCZEK	103
BONCZEK	CARLSON	53
BONCZEK	HUBER	23

CREATION OF RAW COCITATION MATRIX

The next step to prepare the SAS input is creating a raw cocitation matrix. Table 3.3 shows an example of diagonal cell value adjusted cocitation matrix in Microsoft Excel format. The system output can be imported into Microsoft Excel program and need tedious manual "cut and paste" operations to create raw cocitation matrix in the shape of a triangle.

The example of the cocitation matrix data used in this chapter is taken from the actual data used in (Eom 1995; Eom et al. 1996). The data were gathered from a total 692 citing articles in the decision support systems area over the period of 1971-1990. The bibliographic database used here consists of 15,030 cited reference records. Our data consist of only 9 authors throughout this book to help the beginners understand the ACA process more thoroughly and clearly. Chapter 7 provides additional discussions to deal with the issues and problems in dealing with a large number of authors to help beginners in the whole process of author cocitation analysis study.

CONVERSION OF RAW COCITATION MATRIX

It is not suitable to use the raw citation matrix in triangle shape an input to the SAS system for ACA study. The next step is adjusting diagonal cell values in the raw cocitation matrix created.

ADJUSTING DIAGONAL CELL VALUES

Raw cocitation frequencies of row and column authors fill the off-diagonal cells. The off-diagonal cell value is the total number of the cocitation count between these two authors. The term "author" in author cocitation analysis is neither an individual nor individuals. It refers to a body of writings by a person.

The values of the diagonal cells are computed using the adjusted value approach, taking the three highest intersections for each author and dividing the sum by two. The value produced by the cocitation count generation system is to be replaced by the adjusted diagonal cell value. The rationale for using this adjusted value can be found in McCain (1990) In addition to the adjusted diagonal cell value approach, McCain (1990 p.435) discusses the second and third approaches--substituting the diagonal value with the highest off-diagonal cocitation counts for each author and treating the diagonal cell values as missing data. Her initial results indicate little difference between these two. We also found little difference between the approach used here and the second approach of using the highest off-diagonal cell value. The lower part of Table 3.2 shows several rows to illustrate the process of computing the value of each diagonal cell value. In order to prepare a SAS input file, a diagonal cell value adjusted cocitation matrix (Table 3.2) needs further transformation (Table 3.3 Transposed co-citation matrix).

Table 3.2 Diagonal Cell Value Adjusted Cocitation Matrix (1970-1990)

		1	2	3	4	5	6	7	8	9
ALTER	1	121								
BONCZEK	2	25	103							
CARLSON	3	66	53	173						
HUBER	4	23	23	34	68					
KEEN	5	92	46	112	47	206				
SCOTT-MORTON	6	83	42	101	39	174	190			
SIMON	7	38	33	54	37	82	82	111		
SPRAGUE	8	67	59	133	49	126	105	58	182	
WHINSTON	9	27	93	54	24	50	45	37	61	104
Author number		1	2	3	4	5	6	7	8	9
The largest value		92	93	133	49	174	174	82	133	93
Second largest value		83	59	112	47	126	105	82	126	61
Third largest value		67	53	101	39	112	101	58	105	54
Sum		242	205	346	135	412	380	222	364	208
Adj. diagonal value		121	102.5	173	68	206	190	111	182	104

TRANSPOSING ADJUSTED DIAGONAL CELL VALUE COCITATION MATRIX

Table 3.3 is the transposed cocitation matrix converted from Table 3.2. The SAS system does not take the triangle shape data format.

Table 3.3 Transposed Cocitation Matrix (1970-1990)

		1	2	3	4	5	6	7	8	9
ALTER	1	121	25	66	23	92	83	38	67	27
BONCZEK	2	25	103	53	23	46	42	33	59	93
CARLSON	3	66	53	173	34	112	101	54	133	54
HUBER	4	23	23	34	68	47	39	37	49	24
KEEN	5	92	46	112	47	206	174	82	126	50
SCOTT-MORTON	6	83	42	101	39	174	190	82	105	45
SIMON	7	38	33	54	37	82	82	111	58	37
SPRAGUE	8	67	59	133	49	126	105	58	182	61
WHINSTON	9	27	93	54	24	50	45	37	61	104
Author number		1	2	3	4	5	6	7	8	9

A HYPOTHETICAL CASE OF A SIMPLIFIED COCITATION MATRIX

To better understand the raw cocitation matrix used in this study, let's closely examine Table 3.5 using a hypothetical case of a simplified cocitation matrix drawn from three citing reference papers which contain a total of 15 cited references as shown in Table 3.4.

Table 3.4 Reference of Citing Papers

Refer. of Paper #1	Refer. of Paper #2	Refer. of Paper #3
Ackoff	Ackoff	Ackoff
Bonczek	Ackoff	Ackoff
Bonczek	Applegate	Blanning
Blanning	Applegate	
Blanning	Whinston	
Blanning		
Whinston		

Table 3.5 Sample Cocitation Matrix

	Ackoff	Applegate	Bonczek	Blanning	Whinston
Ackoff	<u>2.5</u>				
Applegate	1	<u>1</u>			
Bonczek	1	0	<u>2</u>		
Blanning	2	0	2	<u>3</u>	
Whinston	2	1	1	2	<u>2.5</u>

Each cell value in the cocitation matrix refers to the cocitation counts of paired authors. Raw cocitation frequencies of row and column authors fill the off-diagonal cells.

<u>CONVERSION OF RAW COCITATION MATRIX WITH ADJUSTED VALUES IN DIAGONAL VALUES</u>

There is a difference in computing frequencies between the off-diagonal cells and the diagonal cells. The off-diagonal cells are filled with raw cocitation frequencies of row and column authors. The cell value 1 in Table 3.5, an intersection of the first column (Ackoff) and the second row (Applegate), is the total number of the cocitation count between these two authors that appeared in the citing papers 2 as shown in Table 3.4. Again, the term "author" in author cocitation analysis is neither an individual nor individuals, but a body of writings by a person. Applying that definition to Table 3.4, notice that the intersecting cell between the Ackoff column and the Applegate row in Table 3.5 contains only value '1' to indicate the frequency with which any work by Ackoff is linked to any work by Applegate, despite the fact that the citing paper #2 contains two articles by each.

The diagonal value is computed by summing all cocitation rates in the off-diagonal cells for each author in Table 3.5 and dividing by the total number of authors minus 1 (to exclude the author himself or herself). The mean cocitation rate of Ackoff and Whinston are 1.5 and 0.5, respectively. The diagonal cells are indicated by __. The value of the diagonal cells is computed using the adjusted value approach, taking the three highest intersections for each author and dividing the sum by two.

CHAPTER 4 THE FACTOR PROCEDURE

This chapter describes the factor procedure. The first section of the chapter explains why ACA uses factor analysis. In addition, the chapter discusses the following topics.

- Overview of Input, Procedures, and Outputs of ACA SAS Analysis
- Definition of Factor Analysis
- Getting the Data into a SAS Data Set
- Preparing an ACA Proc Factor Program
- Processing the Inputs
- Interpreting the Factor Procedure Outputs

Numerous multivariate analysis tools exist. The appropriate research methods are determined by three important questions regarding the characteristics of the variables under study (Cooper et al. 1995, p. 521).

1) Are there dependent variables in the problems?
2) Is there more than one dependent variable?
3) Are the variables metric or nonmetric?

Variables in author cocitation analysis are authors. The author is defined as "a body of writings by a person" or "a body of contributions by a person." The term "a person" refers to a single author or one of multiple authors. Authors are not dependent on each other, although one author may exercise some influence on other authors. The second important question is whether the variables are metric (quantitative) or nonmetric (qualitative). Metric variables are measured by ratio and interval measurements; nonmetric variables refer to data that are nominal or ordinal. Nominal data indicate classes and categories not measurable by the quantitative units such as kg, meter, miles, pounds, frequency, etc. Examples of nominal data include an individual's religion, nationality, gender, marital status, union membership status, etc. Since our data has no dependent variables and is metric data, three multivariate analysis tools are applied in this research: factor analysis, cluster analysis and multidimensional scaling.

All three techniques used in the ACA aim at grouping/classifying all variables into several subgroups with common underlying hidden structures, characteristics and/or attributes. The hidden structures/characteristics/attributes are given different terms: factors in factor analysis, clusters in cluster analysis, and dimensions in multidimensional scaling. Although all three techniques seek to summarize/simplify a large number of variables, there are some distinctive differences among these techniques. The basic concepts and some differences among the three techniques are discussed later in chapters 4, 5, and 6.

OVERVIEW OF INPUT, PROCEDURES, AND OUTPUTS

As Figure 4.1 shows, ACA data analysis requires the following 7 different procedures of the SAS system.

- Factor Procedure
- Xmacro Procedure
- Distnew Procedure
- Cluster Procedure
- MDS Procedure
- Plot Procedure

- G3D Procedure

Each procedure requires inputs to produce outputs. It is extremely important for ACA researchers to prepare input files to these procedures in a most efficient manner. There is one necessary input (cocitation frequency matrix), seven procedures (Factor, Xmacro, distnew, cluster, MDS, PLOT, and G3D), and several outputs from 3 multivariate statistical procedures. Factor procedures produce three outputs-- rotated factor pattern, factor structure correlation, and interfactor correlation. Cluster procedures generate cluster history and dendrogram (tree graph). Multidimensional scaling procedures result in iteration history, convergence status, coordinate matrix of each author (first column), and configuration of each author on multidimensional spaces (dimension 1-column 2, dimension 2-column 3, dimension 3-column 4, etc).

Figure 4.1 Factor Procedure in ACA

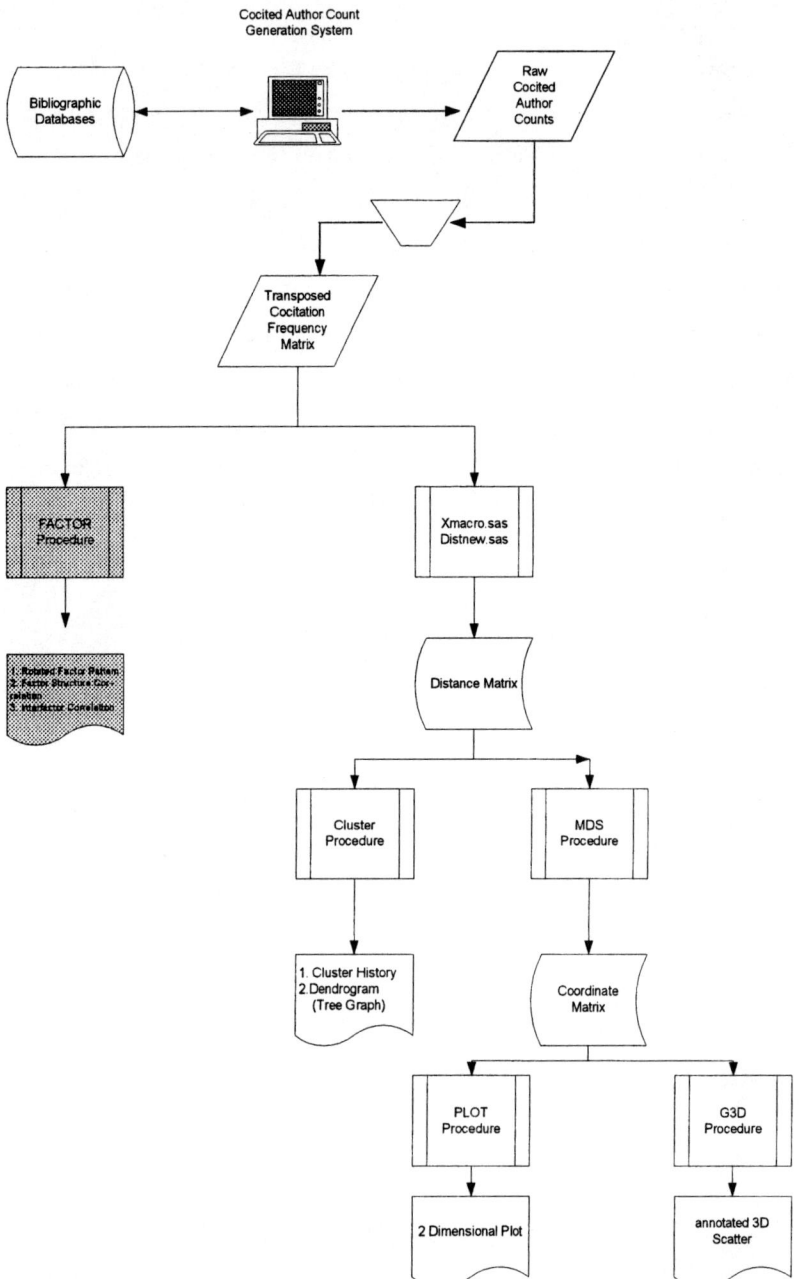

DEFINITION OF FACTOR ANALYSIS

Factor analysis is defined (Kim et al. 1978) as "a variety of statistical techniques whose common objective is to represent a set of variables in terms of a smaller number of hypothetical variables." For those who are not familiar with this technique, see (Child 1970; Hair et al. 1987; Kim et al. 1978) for the basics of factor analysis.

The transposed, diagonal-cell-adjusted cocitation matrix of authors (see Table 4.1) can be analyzed by the factor analysis program in most popular statistical software such as SAS (statistical analysis systems) or SPSS to ascertain the underlying intellectual structure of an academic discipline. The objective of this analysis is to group (condense) a large number of selected variables (authors) into a smaller set of composite dimensions (factors) representing research subspecialties/subdisciplines and contributing disciplines of a discipline. In doing so, each variable (author) is viewed as a dependent variable that is a function of a set of latent (underlying) factors.

Factor loadings represent the correlations between the original variables and the underlying factors. An eigenvalue (latent root or extracted variance) represents the amount of common variance accounted by a factor. The percentage of total variation is an index used to determine how the variables in each factor are related to each other. A high percentage variance means a high degree of interrelatedness among the variables with significant loadings.

GETTING THE DATA INTO A SAS DATA SET

Understanding how the data is inputted to SAS program is crucial to become a good SAS user. Throughout this book, the primary reference to SAS programming is SAS Online Documentation Version 8 (SAS Institute Inc. 2000). Even if I fail to give proper credits at all times, all credits go to *SAS Onlinedoc®* and other SAS publications.

Basically there are two approaches in preparing SAS data files. The first approach includes all statements in one file without using in-file statements and/or permanent output files generated from previous steps. This approach creates a bigger file than the second approach.

MANUAL DATA INPUTS WITHOUT IN-FILE STATEMENTS OR PERMANENT DATA SETS

Depending on the statistical packages used, input formats will be different. Transposed cocitation matrix (Table 3.3) needs further modification to be an input to SAS factor procedures as described below.

First, the first row of Table 3.3, representing author numbers, must be removed. The column names are added to help readers understand better in preparing Table 3.4. The file type must be changed from the spreadsheet file to MS-DOS text file. Spreadsheet programs such as Microsoft Excel are good for building and transforming author cocitation frequency matrix, SAS typically process MS-DOS text formatted datasets. Microsoft Excel, for example, offers several "save as" options including several versions of Microsoft Excel workbook (*.xls), web page (*.htm; .html), comma delimited CSV (*.csv), and several text file options. The text files can be created by different options such as MS-DOS text, Macintosh text, formatted text (space delimited), tab-delimited text, and Unicode text. Out of these options, MS-DOS text file is the right format for preparing SAS input files. To do so, open the transposed cocitation matrix (e.g., Table 3.3) with Excel and save it as MS-DOS text. Reading this MS-DOS formatted file into SAS program will result in a display of all data items of each author on one line.

The SAS input area will display the data set of 9 columns and 9 rows (Figure 4.4). If the data set to be processed is very large, the number of columns will be truncated. In preventing the data sets from being truncated, hard returns must be manually added at the end of each data line. For example, if an author's cocitation frequencies consist of three lines of 36 data points (12 on each line), a hard return must be inserted by pressing the enter key at the end of each of the three lines.

Table 4.1 Creating a Temporary Data Set with Variable Number

```
DATA table41;
INPUT Author $ 1-12 @15 x1-x9;
CARDS;
ALTER           121    25    66    23    92    83    38    67    27
BONCZEK          25   103    53    23    46    42    33    59    93
CARLSON          66    53   173    34   112   101    54   133    54
HUBER            23    23    34    68    47    39    37    49    24
KEEN             92    46   112    47   206   174    82   126    50
SCOTT MORTON     83    42   101    39   174   190    82   105    45
SIMON            38    33    54    37    82    82   111    58    37
SPRAGUE          67    59   133    49   126   105    58   182    61
WHINSTON         27    93    54    24    50    45    37    61   104
;
PROC PRINT;
RUN;
```

Second, several SAS statements must be added to the transposed cocitation matrix file as shown in Table 4.1. The SAS program consists of the following statements.

- Data statements
- Input statements
- Cards statements
- Proc statements
- Run statements

To enter data directly or copy and paste into the SAS system, SAS statements use the DATA, INPUT, and CARDS as shown in Table 4.1.

THE DATA STATEMENT

The first line of the data file (Table 4.1) begins with the DATA statement. General form is DATA data-set-name, ending with semi-colon (;). The data set in Table 4.1 is named table41. Of course, other names are acceptable such as ACAauthor, D1, Dataone, etc. The data set name begins with a character, and no special characters or blank spaces are allowed. It can have as many as eight characters.

THE INPUT STATEMENT

Line 2 begins with the INPUT statement. It provides a name for the variables to be included in the data set. The INPUT statement has the following general syntax form.

INPUT Variable $ beginning column-number –
ending column-number #line-number
    @column-number (variable-name)
    @column-number (variable-name)
    @column-number (variable-name)
    @column-number (variable-name)

- Variable-name is to be displayed as part of outputs. Authors names in the cards statement will not be part of the SAS output. The variable names defined in the INPUT statement (variable-name) are to be printed on the SAS outputs. Therefore, the SAS outputs include author variables as x1, x2, ... x9.
- $ indicates to store a variable value as a character value rather than a numeric value.
- The INPUT statement includes two pointer control statements: line pointer and column pointer (#line-number n moves the pointer to line n. @column n moves the pointer to column n.

Table 4.1 includes only the column pointer control statement because the data of each variable can be listed on one line. Table 4.3 shows both the line control and column control statements because each variable has 67 data that can be listed on 5 lines. The first data-line statement (#1) is optional and omitted in the table.

Rather than listing each variable individually, the tables list variables as a group of nine, x1-x9, in Table 4.1 or a group of fifteen, x1-x15, in Table 4.3. The INPUT statement ends with a semicolon. The difference between Tables 4.1 and 4.2 is whether the names of authors are printed as part of the SAS outputs. Table 4.1 produces outputs using numerical variable names, while Table 4.2 prints authors names as part of the SAS outputs. Including authors' name in the INPUT statement as in Table 4.2 is recommended since the outputs with authors' names helps ACA researchers improve the output interpretations in assigning the names of factors.

THE CARDS STATEMENT

The CARDS Statement signals SAS to read the raw data following after the CARDS statement. The raw data begins after the CARDS statement with a semicolon.

The DATA statement signals the SAS system to begin a step to create a SAS data set. The DATA, INPUT, and CARDS statements create a dataset that begins with WORK. Any dataset whose name begins with WORK is a temporary SAS data set, which are automatically deleted at the end of the session. Tables 4.1 through 4.3 show many different ways of creating temporary data sets for further processing. Table 4.1 is to be used as an input to other procedures and saved as a temporary data set.

Table 4.2 Creating a Temporary Data Set with Author Names in the Input Statement

```
DATA table42;
INPUT @3 Alter Bonczek Carlson Huber Keen ScottMorton
Simon Sprague Whinston;
CARDS;
1    121   25    66    23    92    83    38    67    27
2    25    103   53    23    46    42    33    59    93
3    66    53    173   34    112   101   54    133   54
4    23    23    34    68    47    39    37    49    24
5    92    46    112   47    206   174   82    126   50
6    83    42    101   39    174   190   82    105   45
7    38    33    54    37    82    82    111   58    37
8    67    59    133   49    126   105   58    182   61
9    27    93    54    24    50    45    37    61    104
;
PROC PRINT;
RUN;
```

Table 4.3 Creating a Temporary Data Set with Line Pointer Control Statements

```
DATA Table43;
INPUT @6 X1-X15
    #2 @6 X16-X30 #3 @6 X31-X45
    #4 @6 X46-X60 #5 @6 X61-X67;
CARDS;
1    43  10   9   8  12   3   6  10   5  16  10   8   8   2   8
     9   1  16   6   7   6   2   6   7  13   4  10   6  11  13
     4   3   2   1  33   1   6   8  13   5   7   7   2   8  19
     6  14  18   3  11   9   7   2   4   4  27   3  26  13   4
     7   5   8   3   6  11  10

more data for author variable 2 through 66 here

67  10  17   4   3  26   3   4   8   4  20  16  13   8   2   6
    17  19  25   3   6   4   0   2  15  12   2  15   5   8  20
    10   2  10   1  35   0   2  10   7   4   3  18  17   2  19
     4  13  17   2   8   3   1  14   5  10  25   2  21  26   4
     5   5   7   6   7   8  44
;
PROC PRINT;
RUN;
```

DATA INPUTS USING PERMANENT SAS DATA SETS

The second approach is using either in-file statements and/or permanent SAS data sets. In Figure 4.1 inputs to the procedures (e.g., transposed cocitation frequency matrix) and intermediate outputs (e.g., distance matrix, coordinate matrix) can be saved on a hard disk as permanent data sets. This is a modular approach to SAS programming.

The following section explains each of the two options (in-file statements and permanent data sets) for getting the data into the SAS system.

THE INFILE STATEMENT

Rather than repeatedly entering the data into a SAS program, the data is stored a computer; the INFILE statement can be used to tell the SAS system the name and location of the data so that the SAS systems can retrieve the data for processing. The INFILE statement can be a time saving tool for processing a large data set. Prior to using the INFILE statement, the user has to decide the data format and the location of storage for each file for ACA analysis. For this study, the following two files (Tables 4.4 and 4.5) are created and stored on a PC subdirectory. Including names of authors in SAS outputs will facilitate the interpretation of the outputs since the outputs with numerical variable names are very hard to interpret. For that reason, Table 4.4 is used for further processing. Depending on the format of data files, the SAS data statements differ to include the name of authors in the outputs. To show different SAS statements, we created the two data files and saved as ASCII text files with dat extension.

Table 4.4 SAS Data File (9varname.dat) Stored on c:\wp\books\aca\sasdata

ALTER	121	25	66	23	92	83	38	67	27
BONCZEK	25	103	53	23	46	42	33	59	93
CARLSON	66	53	173	34	112	101	54	133	54
HUBER	23	23	34	68	47	39	37	49	24
KEEN	92	46	112	47	206	174	82	126	50
SCOTTMORTON	83	42	101	39	174	190	82	105	45
SIMON	38	33	54	37	82	82	111	58	37
SPRAGUE	67	59	133	49	126	105	58	182	61
WHINSTON	27	93	54	24	50	45	37	61	104

Table 4.5 SAS Data File (9varnoname.dat) Stored on C:\wp\books\aca\sasdata

1	121	25	66	23	92	83	38	67	27
2	25	103	53	23	46	42	33	59	93
3	66	53	173	34	112	101	54	133	54
4	23	23	34	68	47	39	37	49	24
5	92	46	112	47	206	174	82	126	50
6	83	42	101	39	174	190	82	105	45
7	38	33	54	37	82	82	111	58	37
8	67	59	133	49	126	105	58	182	61
9	27	93	54	24	50	45	37	61	104

RETRIEVAL FROM PERMANENT SAS DATA SETS

Data can be retrieved from permanent SAS data sets the user created and saved on the user's computer. As shown in Table 4.6, a permanent data set is created by using LIBNAME statement. The statement tells the SAS system to store the data sets on the directory you designate.

The first line of statement in Table 4.6 specifies the first-level name and location of permanent data file on your computer's hard drive. Creating a permanent data set requires a two parts name such as mysave.auth9, or acadata.author9, etc. The mysave part is the first-level name and auth9 after dot (.) is the second-level name. With permanent data set saved on your hard disk, it is not necessary to create the data set. The users can go directly the factor procedure with the following Proc Factor statement.

```
PROC FACTOR DATA = mysave.auth9 METHOD=principal
MINEIGEN=1 ROTATE=Promax; RUN;
```

Table 4.6 SAS Program Creating a Permanent Dataset

```
DATA mysave.acaname;
INPUT author $ 1-15 @17 X1-X9;
CARDS;
ALTER           121    25   66   23    92    83   38   67   27
BONCZEK          25   103   53   23    46    42   33   59   93
CARLSON          66    53  173   34   112   101   54  133   54
HUBER            23    23   34   68    47    39   37   49   24
KEEN             92    46  112   47   206   174   82  126   50
SCOTT MORTON     83    42  101   39   174   190   82  105   45
SIMON            38    33   54   37    82    82  111   58   37
SPRAGUE          67    59  133   49   126   105   58  182   61
WHINSTON         27    93   54   24    50    45   37   61  104
;
libname mysave 'c:\wp\books\aca\sasdata\perm';
PROC print;
RUN;
```

Table 4.7 Permanent File (mysave.acaname)

```
                      The SAS System         12:30 Saturday, January 18

author         X1    X2   X3   X4    X5    X6   X7   X8   X9

ALTER         121    25   66   23    92    83   38   67   27
BONCZEK        25   103   53   23    46    42   33   59   93
CARLSON        66    53  173   34   112   101   54  133   54
HUBER          23    23   34   68    47    39   37   49   24
KEEN           92    46  112   47   206   174   82  126   50
SCOTT MORTON   83    42  101   39   174   190   82  105   45
SIMON          38    33   54   37    82    82  111   58   37
SPRAGUE        67    59  133   49   126   105   58  182   61
WHINSTON       27    93   54   24    50    45   37   61  104
```

PROC FACTOR STATEMENT

The last part of the SAS program is the PROC (Procedure) step. This step includes programming statements requesting specific statistical analyses of data. The general statement to perform a principal component analysis is as follows:

    PROC FACTOR < options > ;

There are 9 categories of options available in the PROC FACTOR Statement based on Tasks. They are to specify (1)Datasets, (2)Extract factors and communalities, (3) Data Analysis method, (4)Number of factors, (5)Numerical properties, (6)Rotation method, (7)Displayed outputs, (8)the exclusion of the correlation matrix from the OUTSTAT=data set, and (9) Miscellaneous. The PROC FACTOR statement can be used with a number of options as shown above. There are several important options in analyzing cocitation data.

DATA OPTIONS (DATA=)

    DATA=data-set-name

FACTOR EXTRACTION OPTIONS (METHOD=)

    This option specifies the method for extracting factors. The default is METHOD=PRINCIPAL. The two most frequently used factor extraction methods are principal component analysis (or simply component analysis) and common factor analysis. The principal component analysis is used to summarize most of the original information in a minimum number of factors. ACA research uses the principal component analysis to identify the intellectual structure of an academic discipline.

SPECIFYING NUMBER OF FACTORS OPTIONS (MINEIGEN=, NFACTORS=)

MINEIGEN=: Eigenvalue is the column sum of squares for a factor. It represents the amounts of variance accounted by a factor. Minimum eigenvalue criterion is one of several criteria used for the number of factors to be extracted such as latent root (eigenvalue) criterion, a priori criterion, percentage of variance criterion, and scree test criterion. Readers are referred to (Hair et al. 1987) for the details of these criteria. The latent root criterion is the most commonly used technique and simple to apply. According to this criterion, only the factors that have eigenvalues greater than one are considered significant. Otherwise, the analyst may interpret the factors with less than 1 eigenvalues to be insignificant and exclude them for further interpretation. To do so, the analyst may specify MINEIGEN=1.

NFACT=: There is no exact quantitative basis for deciding the number of factors to extract as the final solution. In addition to these two criteria (the scree test and the minimum eigenvalue criteria), another important criterion is the meaningfulness of the factor. Very often, the factors considered to be acceptable under the eigenvalue and/or scree test criteria fails to be included in the final solution because it is difficult to assign a useful name to the factor that represents the common characteristics of all authors under that factor. Thus, it is very important to include the option of specifying the number of factors to extract.

This option specifies the number of factors to be included in the output. N is the number of factors to extract.
NFACT=n

SPECIFYING ROTATION METHOD OPTION

The rotation of factors is a procedure used to achieve a more meaningful factor solution. In most cases, rotation of the factors improves the interpretation of the factor loadings obtained from the same data by reducing some of the ambiguity of an initial un-rotated factor pattern. The two most frequently used factor rotation options are orthogonal and oblique. When rotating the factors using the orthogonal method, the axes are maintained at 90 degrees (right angles). Therefore, the factors are mathematically independent and they are assumed to be uncorrelated with each other. On the other hand, the oblique rotation method assumes that the factors are correlated with each other. Consequently, the oblique solution provides us with the inter-factor correlation information among the factors.

The PROMAX rotation specification (ROTATE = PROMAX) provides both orthogonal and oblique rotations with only one invocation of PROC FACTOR. Out of the two major rotation options, most ACA studies use an oblique rotation method. Compared to an orthogonal rotation method, the oblique factor rotation is "more desirable because it is theoretically and empirically more realistic"(Hair et al. 1987). It allows a more natural rotation without the imposition of orthogonal factors. If the orthogonal rotation method is desirable, the "ROTATE = VARIMAX" option can be specified.

DISPLAYING OUTPUT OPTIONS (SCREE)

The SCREE option produces a plot which displays the size of the eigenvalues associated with each factor. The scree tail test is an approach that determines the optimum number of factors that can be extracted before the amount of unique variance begins to dominate the common variance structure (Cattell 1966). The scree test involves the plotting of the latent roots (eigenvalues) against the number of factors in their order of extraction. To decide the maximum number of factors extracted, we need to connect factor 1 (at the upper left hand side) and factor 2 (at the middle left hand side) using a straight line, then connect factor 2 and factor 3. This process continues until all factors are connected.

EXAMPLES OF ACA PROC FACTOR PROGRAMS

MANUAL DATA INPUTS WITH AUTHOR'S NAMES

One important statement that saves ACA researchers time when interpreting the output of multivariate statistical techniques is the inclusion of author's names. When interpreting the outcomes of factor analysis as shown in a later chapter, it is important to include names of authors in the INPUT statement instead of observation numbers or variable numbers. Including authors' names in the CARDS statement will not print the authors' name on the SAS outputs. The column 1 of the card statement represents variable numbers.

Table 4.8 PROC FACTOR SAS Program with Author's Names

```
DATA table49;
INPUT @3 Alter Bonczek Carlson Huber Keen ScottMorton
Simon Sprague Whinston
;
CARDS;
1    121   25    66    23    92    83    38    67    27
2    25    103   53    23    46    42    33    59    93
3    66    53    173   34    112   101   54    133   54
4    23    23    34    68    47    39    37    49    24
5    92    46    112   47    206   174   82    126   50
6    83    42    101   39    174   190   82    105   45
7    38    33    54    37    82    82    111   58    37
8    67    59    133   49    126   105   58    182   61
9    27    93    54    24    50    45    37    61    104
;
PROC FACTOR METHOD=PRINCIPAL MINEIGEN=1 ROTATE=PROMAX;
RUN;
```

WITH AN EMBEDDED IN-FILE STATEMENT

An embedded in-file statement can be used to prepare a SAS Program. The infile statement assumes that the transposed cocitation matrix data file (9varname.dat) is stored at c:\wp\books\aca\sasdata\9varname.dat. The data file does not contain any of SAS statements including semi-colon (;). When comparing Tables 4.1 and the following SAS program with the INFILE statement, the use of the in-file statement eliminates the CARDS statement and the INPUT statement will be placed after the infile statement.

> **DATA** table49;
> **INFILE** 'c:\wp\books\aca\sasdata\9varname.dat';
> **INPUT** Author $1-15 @17 x1-x9;
> **PROC FACTOR** method=principal mineigen=1 scree
> Rotate=promax;
> **RUN**;

Table 4.9 PROC FACTOR SAS Program using Author's Name with an INFILE Statement

```
DATA table410;
INFILE 'C:\WP\BOOKS\ACA\SASDATA\9varnoname.DAT';
INPUT @3 Alter Bonczek Carlson Huber Keen ScottMorton
Simon Sprague Whinston;
PROC FACTOR METHOD=PRINCIPAL MINEIGEN=1 ROTATE=PROMAX;
RUN;
```

Table 4.10 Factor Procedure using a Permanent data Set

```
libname mysave 'c:\wp\books\aca\sasdata\perm';
PROC FACTOR data=mysave.acaname METHOD=PRINCIPAL MINEIGEN=1
ROTATE=PROMAX;
RUN;
```

Processing the Inputs

The SAS input files are MS-DOS text files that include cocitation frequency data and SAS commands. The input file must be processed. To do so, invoke the SAS system for Windows. Three windows that make up the SAS display manager system (Release 8.01) will be visible. The windows consist of the Results Window at the left hand side, the Log Windows at the top, right-hand side, and the Editor Window at the bottom, right-hand side. SAS input files are placed in the editor window for processing.

The first step is entering the SAS input files in the Editor window area. To do so, click the rectangle shaped icon at the left hand side of "Editor" title area. A drop-down menu appears showing a submenu of Menu, Move, Size, etc. Click the Menu, File, and Open. The Open window will appear on the screen. One can now choose the subdirectory where his or her SAS input file is stored. Figure 4.3 shows the SAS Display Manager System Screen with the Editor Window filled with the input file, fac9INPUTname.sas.

The output to be discussed in this chapter is based on the SAS input file of 9 authors. All the SAS input files discussed in this chapter includes only 9 authors for the sake of simplicity. For complete cocitation matrix data, refer to Eom (Eom 1995). This cocitation matrix is compiled from a total of 692 citing articles in the decision support systems (DSS) area over the past 20 years (1969-1990). For a detailed discussion on these datasets, see Eom (Eom 1995; Eom et al. 1996). To execute the SAS input files entered, click the "Run" menu, followed by "Submit" menu. After the program has run, the SAS display manager shows the Output screen with minimized "results", "data", and "log" screen buttons.

There is a great chance of making errors when learning to use the SAS system. The LOG window displays important information including error messages, number of observations, number of variables, the amounts of processing time used by the SAS system, number of factor to be retained by the NFACTOR criterion.

Figure 4.2 The SAS Display Manager System Screen (Release 8.01)

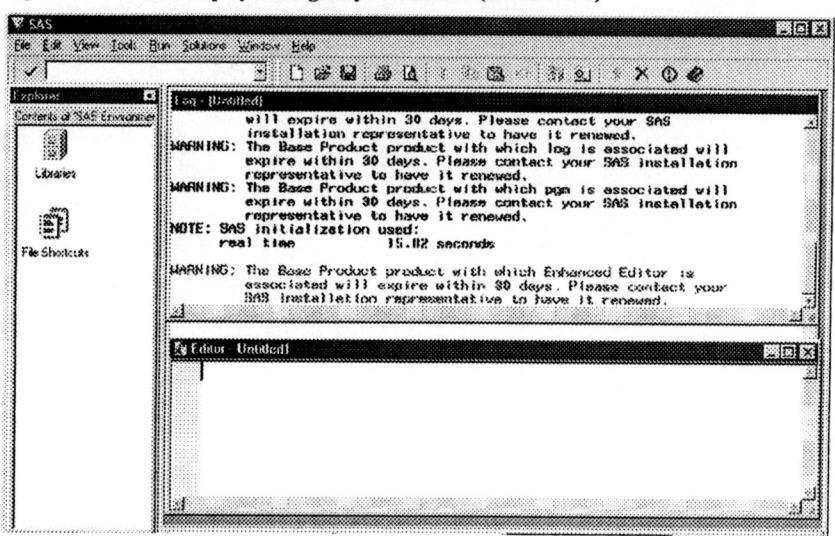

Figure 4.3 The SAS Display Manager System Screen with Factor Analysis Data and Corresponding Outputs (Release 8.01)

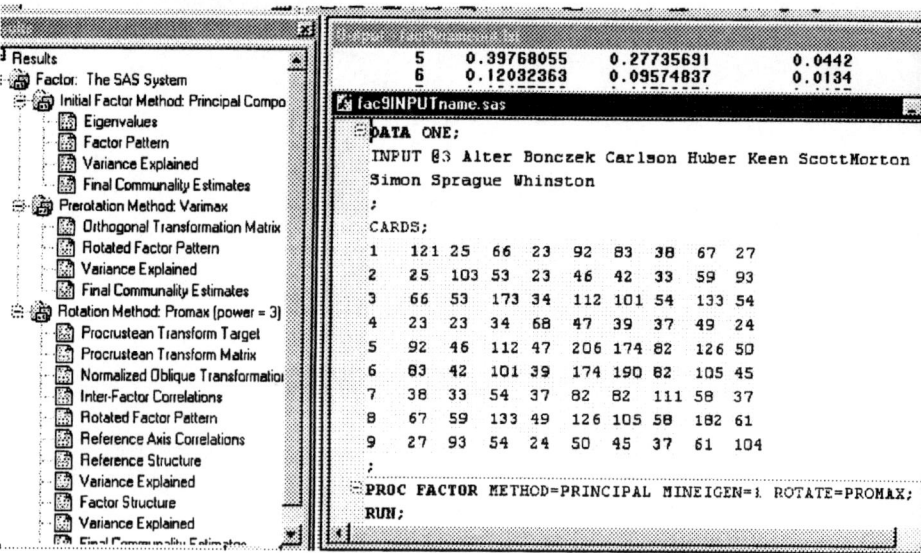

INTERPRETING THE FACTOR PROCEDURE OUTPUTS

As shown in Figure 4.3, the SAS factor procedure produced 19 different outputs under three major headings: initial factor method: principal components; prerotation method: varimax; and rotation method: promax.

INITIAL FACTOR METHOD: PRINCIPAL COMPONENTS

The transposed cocitation matrix of 9 authors is analyzed by the principal components analysis with the latent root criterion (eigenvalue 1 criterion) applied to obtain the initial solution of 4 factors.

The first segments of the output are as follows. ACA analysis uses factor analysis to reduce the number of variables into smaller number of observed variables that tend to hang together. The SAS FACTOR procedure performs both principal components analysis and factor analysis. Although these two techniques are similar, they are not identical due to the assumptions of an underlying causal structure. While principal components analysis assumes no underlying causal structure between the observed variables and latent variables, factor analysis is based on the assumption that "the covariation in the observed variables is due to the presence of one or more latent variables (factors) that exert causal influence on these observed variables."(Hatcher 1994, p.9)

Table 4.11 Initial Factor Method: Principal Components

The SAS System 11:46 Friday, January 25, 2002
The FACTOR Procedure Initial Factor Method: Principal Components
Prior Communality Estimates: ONE
Eigenvalues of the Correlation Matrix: Total = 9 Average = 1

	Eigenvalue	Difference	Proportion	Cumulative
1	4.22575393	2.01474495	0.4695	0.4695
2	2.21100899	1.13125294	0.2457	0.7152
3	1.07975604	0.15566117	0.1200	0.8352
4	0.92409487	0.52641432	0.1027	0.9378
5	0.39768055	0.27735691	0.0442	0.9820
6	0.12032363	0.09574837	0.0134	0.9954
7	0.02457526	0.00776854	0.0027	0.9981
8	0.01680672	0.01680672	0.0019	1.0000
9	0.00000000		0.0000	1.0000

3 factors will be retained by the MINEIGEN criterion.

Factor Pattern

	Factor1	Factor2	Factor3
Alter	0.73621	0.07271	-0.53374
Bonczek	-0.58395	0.77515	0.08406
Carlson	0.66579	0.55143	0.24243
Huber	0.32740	-0.54845	0.73192
Keen	0.92069	0.23074	-0.05239
Scott Morton	0.89789	0.19807	-0.12872
Simon	0.59921	-0.15880	-0.05081
Sprague	0.68328	0.51844	0.40691
Whinston	-0.55898	0.78327	0.07652

Variance Explained by Each Factor

Factor1	Factor2	Factor3
4.2257539	2.2110090	1.0797560

Final Communality Estimates: Total = 7.516519

Alter	Bonczek	Carlson	Huber	Keen
0.83216903	0.94893043	0.80613171	0.94369172	0.90365786

Scott Morton	Simon	Sprague	Whinston
0.86201759	0.38685474	0.90123407	0.93183182

Outputs under the heading of "initial factor method: Principal components" list four elements of information that can be summarized as in Table 4.12. Eigenvalue (the latent root) of a factor is the column sum of squares for that factor. For example, eigenvalue of factor 1 is the sum of squared factor loadings of all variables on that factor and is computed as follows.

$0.73621^2 + (-0.58395)^2 + 0.66579^2 + 0.3274^2 + 0.92069^2 + 0.89789^2 + 0.59921^2 + 0.68328^2 + (-0.55898)^2 = 4.225729$

Factor pattern is a matrix of factor loadings (factor pattern) consisting of three columns of factors 1, 2, and 3 and nine author variable rows. Factor loadings represent the correlation between original author variables and the factors. Variance explained by each factor is identical to the eigenvalue of each factor, often expressed as percentage. The last row of Table 4.12 shows the variance of each factor in terms of percentage, which is the variance of each factor divided by the number of authors (variables). For example, the variance explained by factor 1 (46.95%) is computed as follows.

$(4.225729/9)*100 = 46.9525\%$

Final communality estimates of each author (observed variable) are the sum of the square of factor loadings for that variable. For example, communality for Alter is computed as the sum of the square of each of three factor loadings for that author.

$(0.736212^2 + 0.072712^2 + (-0.53374)^2 =$
$.542005+.005287+.284878=.83217.$

Table 4.12 A Matrix of Factor Loadings with Eigenvalue and Communality

	Factor1	Factor2	Factor3	Communality
Alter	0.73621	0.07271	-0.53374	0.8321703
Bonczek	-0.58395	0.77515	0.08406	0.94892121
Carlson	0.66579	0.55143	0.24243	0.80612367
Huber	0.3274	-0.54845	0.73192	0.94369505
Keen	**0.92069**	0.23074	-0.05239	0.90365574
Scott Morton	**0.89789**	0.19807	-0.12872	0.86200702
Simon	0.59921	-0.1588	-0.05081	0.38685172
Sprague	0.68328	0.51844	0.40691	0.90122734
Whinston	-0.55898	0.78327	0.07652	0.93182584
Eigenvalue	4.225729	2.210999	1.07975	7.51647788
Variance %	46.95%	24.56%	11.99%	83.51%

OPTIMAL NUMBER OF FACTORS

Before proceeding to the next section of output (Prerotation Method: Varimax), it is very important to understand the different approaches to find the optimal number of factors in factor analysis. There are three commonly used.

Eigenvalue one criterion: In Table 4.12, the first column represents the number of principal components extracted from 9 authors (variables). Eigenvalue associated with each variable is listed under the second column with heading "Eigenvalue". Notice that the first component's eigenvalue (4.22) accounts for 46.95% of the total variance. Each of succeeding component accounts for progressively smaller amounts of variance.

From the first part of output (Table 4.12), ACA researchers derive a very important piece of information -- "3 factors will be retained by the MINEIGEN criterion." When preparing input file, ACA researchers begin with no idea with regard to the number of factors to be derived. Using the eigenvalue = 1 criterion, three factor solution is a starting point for finding the meaningful number of factors. The meaningful solution may consist of more or less factors than the three-factor solution. To find the meaningful final solution, ACA researchers need to carefully interpret all solutions to ascertain the meaning of each factor. At this stage, there is no need to examine all factors. Rather, attention should be given to the last one or two factors to assess the meaning of the factor. Therefore, ACA analysts may specify NFACT options to include 2, 3, and 4 factor solutions. In ACA of using custom databases, the meaning factor solution occurs usually smaller than the number of factors identified by the Eigenvalue criterion.

PROC FACTOR METHOD=PRINCIPAL NFACT=2 ROTATE=PROMAX;
PROC FACTOR METHOD=PRINCIPAL NFACT=3 ROTATE=PROMAX;
PROC FACTOR METHOD=PRINCIPAL NFACT=4 ROTATE=PROMAX;

Specifying different number of factors may result in different factor loadings. All the factor loadings under each factor of 1, 2, and 3 may be different from 3 factor solution using NFACT=3 and 4 factor solution using NFACT=4 option.

Scree Test: Scree test is another approach for deciding the optimal number of factors. Figure 4.4 shows that the lines connecting the eigenvalues of factors 1 through 9. Notice that as the number of factors increases, the slope of the lines connecting two successive factors becomes more gentle and almost a horizontal line. This example is taken from another dataset in our previous research. Applying the scree test, a seven-factor solution can be the final solution, even though the minimum eigenvalue 1 criterion suggests that nine factors would be acceptable. Needless to say, identifying the intellectual structure of an academic discipline is not a well-structured process.

The subjective approach: the third approach in deciding the optimal number of factors is the subjective judgment-based approach. Factor analysis in ACA should be a supporting tool that must be used with expert judgments in regard to the interpretability of each factor. Specifically, ACA analyst should have a reasonable level of understanding as to what each author in a discipline under investigation has done in the development of that field. The data that can be used in this process include frequently cited articles in the cited bibliographic databases. ACA analysts should make a judgment on each factor regarding the meaning of the factor. This evaluation of the interpretability of each factor begins from the factors with smaller eigenvalues.

Figure 4.4 Scree Plot of Eigenvalues

        The SAS System   14:42 Friday, January 4, 2002
              The FACTOR Procedure
        Initial Factor Method: Principal Components

Scree Plot of Eigenvalues

```
       ‚
    25 ˆ
       ‚
       ‚
       ‚   1
       ‚
       ‚
       ‚
    20 ˆ
       ‚
       ‚
       ‚
       ‚
 E     ‚
 i     ‚
 g  15 ˆ
 e     ‚
 n     ‚
 v     ‚
 a     ‚   2
 l     ‚
 u     ‚
 e  10 ˆ
 s     ‚
       ‚    3
       ‚
       ‚
       ‚
     5 ˆ
       ‚
       ‚    4
       ‚     5
       ‚      6
       ‚      79
       ‚       0124567
     0 ˆ         9012456790124567901245679012456790124567
       Šƒƒƒƒƒƒƒƒƒƒƒƒƒƒƒƒƒƒƒƒƒƒƒƒƒƒƒƒƒƒƒƒƒƒƒƒƒƒƒƒƒƒƒƒƒƒƒƒƒ
         0    10    20    30    40    50    60    70

                          Number
```

PREROTATION METHOD: VARIMAX

The second part of the output under the "Prerotation Method: Varimax" heading in Table 4.13 lists orthogonal transformation matrix, rotated factor pattern, variance explained, and final communality estimates. The second output from the factor procedure is a rotated factor matrix. The rotation of factor matrix makes the factor solution easier to interpret by reducing the ambiguities in interpreting the factor matrix. Readers are referred to chapter 6 of (Hair et al. 1987) to understand the concept of rotation. Comparing Tables 4.12 (unrotated factor matrix) and 4.13 (rotated factor matrix), the rotated factor matrix includes many variables (authors) whose factor loadings are grater than .8.

Factor loadings matrix can be rotated by using an orthogonal or oblique method. A varimax rotation is the most commonly used orthogonal rotation in ACA study, and it maximizes the variance of a column of the factor pattern matrix, not a row of the matrix. The term "orthogonal" involves right (90 degree) angles, while an "oblique" rotation involves acute or obtuse angles. A rotated factor pattern using a varimax rotation is assumed to have no inter-factor correlations. Notice that rotating factor loading matrix changes (1)factor loadings of each author and (2)variance explained by each factor, but it does not change (1)final communality estimate and (2)communality of each variable (author). Rotation of factor loadings matrix does not change the total variance of the data set (83.51 percent), while redistributing the proportion of variance that can be explained by each factor.

ACA researchers seem to agree that research subspecialties in an academic discipline represented by factors are correlated each other. Therefore, the last two tables are usually included as a basis of inferring the intellectual structure of a discipline.

ROTATED FACTOR PATTERN

The next step is deciding which variables to include in the result tables as a basis of inferring the intellectual structure of a discipline. There are three important tables that interest ACA researchers:

1. Rotated factor pattern tables with variance explained to be taken from the "Prerotation Method: Varimax" section of the output.
2. Factor structure (Correlations) with variance explained from the "Rotation Method: Promax (power -3)" section
3. Inter-factor correlations from the "Rotation method: Promax (power -3) section".

When reporting the result, ACA researchers recommend including only variables with factor loadings whose absolute value exceeds ± .40 and any variable whose absolute value is greater than ±.70 should be used to determine the meanings of each factor.

Table 4.13 Rotated Factor Pattern (Varimax)

The SAS System   11:46 Friday, January 25, 2002
The FACTOR Procedure Prerotation Method: Varimax

Orthogonal Transformation Matrix

	1	2	3
1	0.80457	-0.59156	-0.05215
2	0.55076	0.77614	-0.30704
3	0.22210	0.21831	0.95027

Rotated Factor Pattern

	Factor1	Factor2	Factor3
Alter	0.51383	-0.49560	-0.56791
Bonczek	-0.02424	**0.96542**	-0.12766
Carlson	**0.89323**	0.08705	0.02635
Huber	0.12391	-0.45956	**0.84684**
Keen	**0.85621**	-0.37700	-0.16865
Scott Morton	**0.80292**	-0.40553	-0.22996
Simon	0.38336	-0.48882	-0.03078
Sprague	**0.92566**	0.08701	0.19186
Whinston	-0.00135	**0.95531**	-0.13862

Variance Explained by Each Factor

Factor1	Factor2	Factor3
3.4594143	2.8621496	1.1949551

Final Communality Estimates: Total = 7.516519

Alter	Bonczek	Carlson	Huber	Keen
0.83216903	0.94893043	0.80613171	0.94369172	0.90365786

Scott Morton	Simon	Sprague	Whinston
0.86201759	0.38685474	0.90123407	0.93183182

ROTATION METHOD: PROMAX

<u>INTERFACTOR CORRELATIONS</u>

Since the majority of ACA researchers seem to agree that subspecialties in an academic discipline are interrelated, it is not necessary to include the rotated factor pattern tables, but it may help ACA researchers by providing supplementary information.

The PROMAX rotation specification provides both orthogonal and oblique rotations with only one invocation of PROC FACTOR. Compared to an orthogonal rotation method, the oblique factor rotation is "more desirable because it is theoretically and empirically more realistic" (Hair et al. 1987 p. 245). It allows a more natural rotation without the imposition of orthogonal factors. Moreover, it generates additional information about the correlations between the factors (Table 4.14). Oblique rotation method provides important information that cannot be found with orthogonal rotation method. For oblique rotation-based factor solutions, proper interpretation of a set of factors requires examining the factor pattern (the weight matrix to calculate variable standard scores from factor standard scores), the factor structure (the correlations of the variables with the factors), and the reference structures (the correlations between the variables and the factors when the variance attributable to all other factors has been removed). That is the interfactor correlations (Table 4.14). SAS output, interfactor correlations, provides ACA researchers with very useful information about the influence of each factor has on the development of other factors. To assess the impact of one research subspecialty on the other subdisciplines, ACA researchers often construct major factor intercorrelation network such as Figure 7.1. The intercorrelation network shows that organization science, a reference discipline, has made important contributions to the development of decision support systems research subspecialties, especially in the areas of foundations, model management, and user-interfaces. For further details on this subject, refer to (Eom et al. 1996).

Table 4.14 Interfactor Correlations (1970-1990)

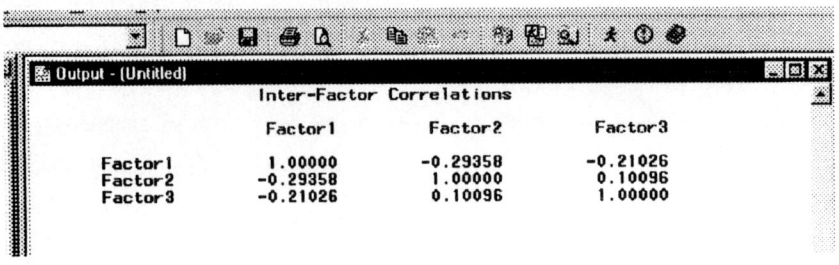

FACTOR STRUCTURE CORRELATIONS

Table 4.15 Factor Structure Correlations

```
                Factor Structure (Correlations)

                    Factor1         Factor2         Factor3
    Alter           0.61546        -0.58566        -0.65800
    Bonczek        -0.15499         0.95336        -0.07645
    Carlson         0.86775        -0.04786        -0.09608
    Huber           0.13411        -0.44518         0.79776
    Keen            0.91071        -0.50712        -0.30594
    ScottMorton     0.86616        -0.52927        -0.36040
    Simon           0.45105        -0.54179        -0.10811
    Sprague         0.88912        -0.04741         0.06299
    Whinston       -0.13023         0.93956        -0.09101

         Variance Explained by Each Factor Ignoring Other Factors

                Factor1         Factor2         Factor3
               3.7643555       3.1682109       1.3318792

         Final Communality Estimates: Total = 7.516519

    Alter       Bonczek       Carlson        Huber         Keen
  0.83216903   0.94893043   0.80613171    0.94369172   0.90365786
```

CHAPTER 5 THE CLUSTER PROCEDURE

This chapter describes the cluster procedure of the SAS system. Cluster analysis is a data reduction technique for grouping various entities (individuals, variables, objects) into clusters so that the entities in the same cluster have more similarity each other with respect to some predetermined selection criteria (Everitt 1980; Hair et al. 1992). Therefore, it is necessary to convert the raw cocitation frequency matrix into a measure of similarity or distance. To do so, SAS institute has developed the DISTANCE macro for computing various measures of distance, dissimilarity, or similarity between the observations of a SAS data set. The first section of this chapter explains the creation of a distance matrix, which is the input to the cluster procedure. The second part of this chapter focuses on the PROC CLUSTER statement which sets out the CLUSTER procedure steps. This chapter includes the discussions of the following topics.

- Generations of a distance matrix
- Creating a permanent distance matrix
- Proc Cluster Statement
- Interpreting Results of Cluster Analysis

GENERATION OF A DISTANCE MATRIX USING XMACRO.SAS AND DISTNEW.SAS

Figure 5.1 shows that two SAS programs (xmacro.sas and distnew.sas) process a transposed cocitation matrix (input) to produce a distance matrix (output). There are many different ways of measuring inter-object similarity, including distance measures (proximity/difference between each pair of objects) and the correlation coefficient between a pair of objects. In ACA, the higher cocitation frequencies between a pair of authors represents a high level of cognitive linkages or similarities between them. The preliminary documentation of the DISTANCE macro is provided by SAS institute as a service to the user (Kuo 1997). Here is the link to the documentation.
http://www.sas.com/service/techsup/faq/stat_macro/distmacr752.html

THE %INCLUDE STATEMENT

The %INCLUDE statement allow a SAS program to bring (1)an entire external file or (2)specifies lines that are entered earlier in the same SAS job or (3)statements or data lines that you enter directly from the terminal into the program. The selective inclusion of lines from an external file is not permitted. Although one %INCLUDE statement may include multiple sources describing the location of the information that the user wants to access, it is easier to understand a program that uses separately codes %INCLUDE statements for each source. Table 5.1 included two %INCLUDE statements to use xmacro.sas program and distnew.sas program. Xmacro.sas is a SAS sample utility library. The macros in xmacro.sas are intended to make it relatively easy to write macros that act much like SAS procedures with respect to error checking, setting defaults, using variable lists, and doing BY processing.

Figure 5.1 The Cluster Procedure in ACA

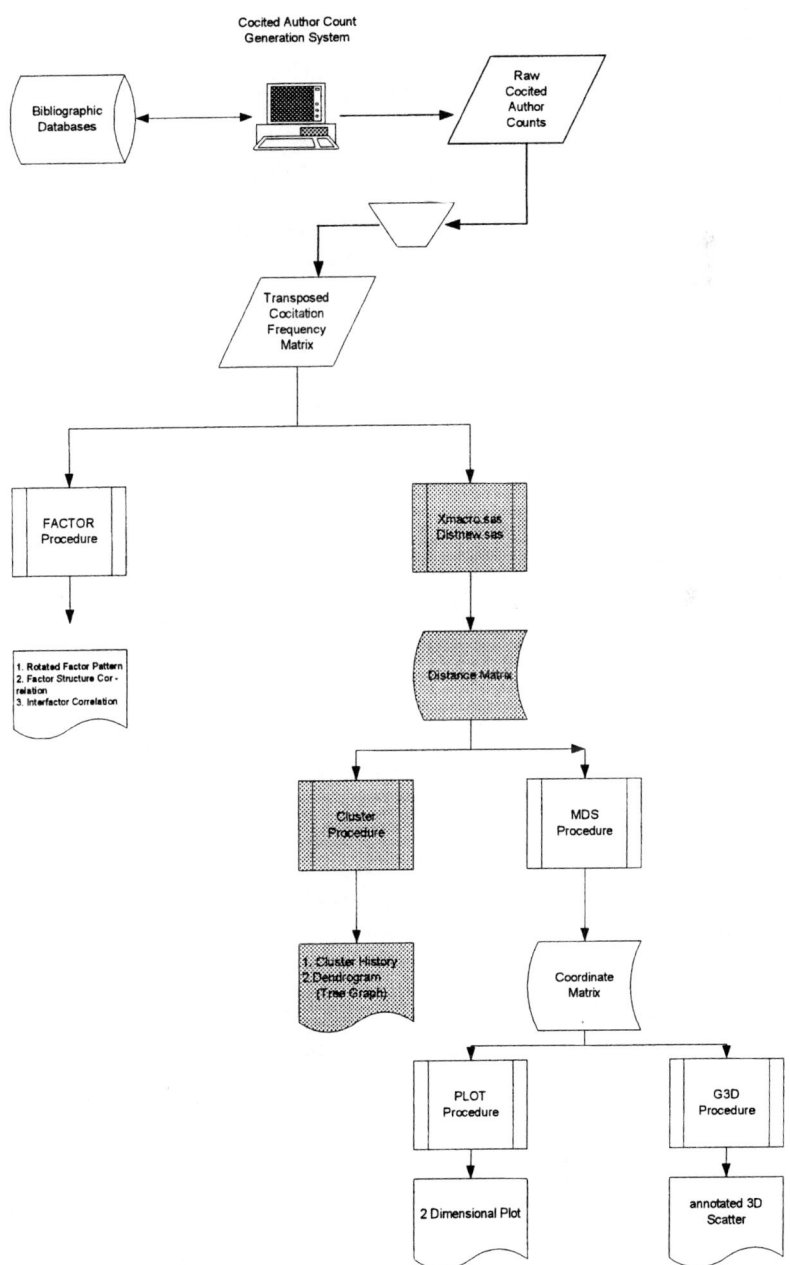

%DISTANCE ARGUMENTS

The %DISTANCE is a macro used to compute the distance between each pair of variables. The %distance macro needs data values and several options. The data values and options can be listed within parentheses. They are called arguments and separated by commas. The following arguments may be listed within parentheses in any order, separated by commas. See (Kuo 1997) for the details of the distance macro and its arguments.

%distance (data=aca9var,
          id=author,
          options=print,
          method=dcorr,
          var=x1-x9,
          out=dist);

SPECIFY THE INPUT DATA SET

    The data option is needed to specify the name of data set. In ACA, the data is a distance matrix among authors converted from the raw cocitation frequency matrix (see Table 5.6).

    Data = SAS-data-set

PRINT THE NAME OF ALL AUTHORS IN THE OUTPUTS

    The ID option allows printing the text values of the variables. The variables in ACA are cocitation frequencies of all authors. Without the ID statement, the output of this distance macro lists variables x1 through x9. The ID statement is used to print the name of all authors in the outputs. This option is an important tool that can facilitate the interpretation of outputs easier and faster.

ID = authors

LIST ADDITIONAL OPTIONS

Additional options are PRINT, NOMISS, REPLACE, and REPONLY.

Options= PRINT

SPECIFY THE METHOD FOR COMPUTING DISTANCE MEASURES

Unless specified otherwise, each method for computing distance or similarity measures allows only numeric variables. The method is one of the following 15 methods.

Method = DCORR

EUCLID	Euclidean distance
SIZE	Size distance
SHAPE	Shape distance
COV	Covariance
CORR	Correlation
DCORR	Correlation transformed to Euclidean distance as sqrt (1-CORR)
SQCORR	Squared correlation
DSQCORR	One minus squared correlation
L(p)	Minkowski L(p) distance, where p is a positive numeric value
SIMRATIO	Similarity ratio (if variables are binary, this is the Jaccard coefficient)
DISRATIO	One minus similarity ratio
JACCARD	Jaccard similarity coefficient computed from numeric variables where 0=absence, 1=presence, and intermediate values are allowed
DJACCARD	Jaccard dissimilarity coefficient
MATCH	Simple matching coefficient; allows mixed variables

DMATCH   Simple matching coefficient transformed to Euclidean distance; allows mixed variables

For use in PROC CLUSTER, the D... transformations of similarity measures should be used; for example, METHOD=DCORR rather than METHOD=CORR. The value of the METHOD= argument is the name of a macro to compute the distance or similarity measure between two vectors. The user can write additional macros to implement other distance or similarity measures besides those listed above.

Euclidian distance is the most commonly used measure of similarity between two objects in two dimensions. Suppose there are point A (X_1, Y_1) and point B(X_2, Y_2) on a two dimensional space (X-axis and Y-axis). The distance is measured by the length of the hypotenuse of a right angle, as calculated by square root of $[(X_2 - X_1)^2 + [(Y_2 - Y_1)^2]$.

SPECIFY THE VARIABLES IN THE INPUT DATA SETS

The VAR statement is used to specify the variables in the input data set that are to be analyzed by the procedure. The variables may be numeric (e.g., x1-x10), character (abc—xyz), etc.

Var = x1-x9

SPECIFY THE OUTPUT DATA SET TO BE PRODUCED

The output data set produced by the OUT= option of the %distance macro. The distance matrix (dist) contains a triangle shape of distance matrix converted from correlations coefficients.

Out = Output data-set name

These proximity measures are stored as a lower triangular matrix in an output data set that can then be used as input to the CLUSTER, multidimensional scaling (MDS), or MODECLUS procedures. The input data set may contain numeric variables, character variables, or both, depending on which proximity measure is used. ACA uses the measurements of inter-objective similarities using correlation transformed to Euclidean distance using METHOD=DCORR option as an input to the PROC CLUSTER procedure. The above SAS statement produces the distance matrix (9 observation rows and 9 distance columns, triangle shape) as shown in Table 5.4.

CREATING THE PERMANENT DISTANCE MATRIX

Prior to conduct cluster analysis, the transposed cocitation frequency matrix must be converted into a distance matrix using Xmacro.sas and distnew.sas program. Table 5.2 shows a SAS program for creating a permanent distance data set for cluster analysis. The data step includes the DATA, INPUT, and CARDS statements. A distance matrix is the input to the cluster procedure as well as MDS procedure. To save time, a distance matrix is created as a permanent data set (Tables 5.1 and 5.2). The SAS sample library now contains a new version of the distance macro, distnew, updated on May 01, 2001. The distance macro requires the XMACRO macro to convert a cocitation frequency matrix into a proximity measure that can be used as the input to the CLUSTER and MDS procedures. Thus, the following statements must be included before the PROC cluster or PROC MDS statement (see Table 5.1).

```
%include 'c:\program files\SAS Institute\SAS\V8\stat\sample\xmacro.sas';
%include 'c:\program files\SAS Institute\SAS\V8\stat\sample\distnew.sas';
```

Table 5.1 Creating Permanent Distance Data Set with Card Input

DATA ACA9VAR;
INPUT Author $ **1-15** @**17** x1-x9;
CARDS;

ALTER	121	25	66	23	92	83	38	67	27
BONCZEK	25	103	53	23	46	42	33	59	93
CARLSON	66	53	173	34	112	101	54	133	54
HUBER	23	23	34	68	47	39	37	49	24
KEEN	92	46	112	47	206	174	82	126	50
SCOTT MORTON	83	42	101	39	174	190	82	105	45
SIMON	38	33	54	37	82	82	111	58	37
SPRAGUE	67	59	133	49	126	105	58	182	61
WHINSTON	27	93	54	24	50	45	37	61	104

;
libname mysave 'c:\wp\books\aca\sasdata\';
%include 'c:\program files\SAS Institute\SAS\V8\stat\sample\xmacro.sas';
%include 'c:\program files\SAS Institute\SAS\V8\stat\sample\distnew.sas';
%distance (data=aca9var,
        id=author,
        options=print,
        method=dcorr,
        var=x1-x9,
           out=mysave.dist);
proc print data=mysave.dist;
run;

Table 5.2 Creating Permanent Distance Data Set with INFILE Statement

DATA ACA9VAR;
infile 'c:\wp\books\aca\sasdata\9varname.dat';
INPUT Author $ **1-15** @**17** x1-x9;
libname mysave 'c:\wp\books\aca\sasdata\perm';
%include 'c:\program files\SAS Institute\SAS\V8\stat\sample\xmacro.sas';
%include 'c:\program files\SAS Institute\SAS\V8\stat\sample\distnew.sas';
%distance (data=aca9var,
        id=author,
        options=print,
        method=dcorr,
        var=x1-x9,
        out=mysave.dist);
proc print data=mysave.dist;
run;

PROC CLUSTER STATEMENT

Three main stages must be followed in applying cluster analysis to any problem. They are partitioning, interpretation, and profiling stages. The partitioning stage is concerned with the separation of a whole data set into several groups (clusters). In this stage, three key questions must be carefully considered (Hair et al. 1992).

> (1) Similarity measures: How should the inter-object similarity be measured?
> (2) Clustering algorithms: What procedure (clustering algorithms) should be used to place similar object into clusters?
> (3) Selecting the number of clusters: How many clusters should be selected?

The basic syntax consists of PROC CLUSTER statement with options. This procedure builds a tree-like structure (dendrogram/tree graph), which can be a very useful tool for understanding the evolution of thought in an academic discipline. The dendrogram shows a chronological order of the emergence of each decision support systems subspecialty as well as their interdependency on one another. While factor analysis output produces a snap-shot of invisible/visible colleagues interacting together, cluster analysis output (dendrogram) shows a historical pattern of interaction/cooperation among the colleagues.

PROC CLUSTER METHOD=NAME <OPTIONS>;

    BY variables;
    COPY variables;
    FREQ variables;
    ID variables;
    RMSSTD variables;
    VAR variables;

The Table 5.4 summarizes the options in the PROC CLUSTER statement (Source: SAS Online Doc, Version 8). Of these options available, the PROC Cluster Statement in Table 5.3 uses only DATA=, METHOD=, Pseudo options, and ID variables.

SPECIFY THE INPUT DATA SET

    DATA= SAS-data-set

SPECIFY CLUSTERING METHODS

    METHOD= {AVERAGE, CENTROID, COMPLETE, DENSITY, EML, FLEXIBLE, MCQUITTY, MEDIAN, SINGLE, TWOSTAGE, WARD}

    The PROC CLUSTER procedure uses 11 clustering algorithms to compute the distance measures between two clusters. The algorithms are AVERAGE, CENTROID, COMPLETE, DENSITY, EML, FLEXIBLE, MCQUITTY, MEDIAN, SINGLE, TWOSTAGE, and WARD. The details of the methods are available at http://v8doc.sas.com/sashtml/. ACA study uses the WARD method.

    Two most commonly used general categories of clustering algorithms are hierarchical and nonhierarchical.

Hierarchical clustering procedures can take either agglomerative or divisive procedures. In ACA research, we are interested in subdividing all authors in a discipline into several groups with similar research subspecialties. The agglomerative procedure clusters each individual author into several groups. The first step in the agglomerative clustering procedure starts with combing the two objects which have the smallest distance measure.

In Ward's minimum-variance method, the distance between two clusters is the *ANOVA* sum of squares between the two clusters added up over all the variables. At each generation, the within-cluster sum of squares is minimized over all partitions obtainable by merging two clusters from the previous generation. The sums of squares are easier to interpret when they are divided by the total sum of squares to give proportions of variance (squared semipartial correlations).(SAS Institute Inc. 1988).

All these procedures maximize the differences between clusters relative to the variations within clusters. The ratio of the between-cluster variation to the average within-cluster variation is then comparable to the F-ratio in analysis of variance (Hair et al. 1992).

CONTROL DISPLAY OF THE CLUSTER HISTORY

PSEUDO is an option to print pseudo F and t^2 statistics. This option is effective only when the data are coordinates or METHOD=AVERAGE, CENTEROID, OR WARD.

ID VARIABLES

With the ID statement, cluster history and tree graph will use name of authors instead of OBn, where n is the observation number.

Table 5.3 shows how to perform the cluster procedure using the permanent data set (mysave.dist).

Table 5.3 PROC CLUSTER SAS Statement with a Permanent Data Set

libname mysave 'c:\wp\books\aca\sasdata\perm';
PROC CLUSTER METHOD=WARD DATA= mysave.dist PSEUDO;
ID author;
PROC TREE SPACE=2;
RUN;

Table 5.4 Options in the PROC CLUSTER Statement

Tasks	Options
Specify input and output data sets	
specify input data set	DATA=
Create output data set	OUTTREE=
Specify clustering methods	
specify clustering method	METHOD=
beta for flexible beta method	BETA=
minimum number of members for modal clusters	MODE=
penalty coefficient for maximum-likelihood	PENALTY=
Wong's hybrid clustering method	HYBRID
Control data processing prior to clustering	
suppress computation of eigenvalues	NOEIGEN
suppress normalizing of distances	NONORM
suppress squaring of distances	NOSQUARE
standardize variables	STANDARD
omit points with low probability densities	TRIM=
Control density estimation	
Dimensionality for estimates	DIM=
number of neighbors for kth-nearest-neighbor	K=
Radius of sphere of support for uniform-kernel	R=
Suppress checking for ties	NOTIE
Control display of the cluster history	
display cubic clustering criterion	CCC
suppress display of ID values	NOID
specify number of generations to display	PRINT=
display pseudo F and t^2 statistics	PSEUDO
display root-mean-square standard deviation	RMSSTD
display R^2 and semipartial R^2	RSQUARE
Control other aspects of output	
suppress display of all output	NOPRINT
display simple summary statistics	SIMPLE

Table 5.5 PROC CLUSTER Statement

```
DATA ACA9VAR;
INPUT Author $ 1-12 @13 x1-x9;
CARDS;
ALTER         121   25   66   23    92   83   38    67   27
BONCZEK        25  103   53   23    46   42   33    59   93
CARLSON        66   53  173   34   112  101   54   133   54
HUBER          23   23   34   68    47   39   37    49   24
KEEN           92   46  112   47   206  174   82   126   50
SCOTTMORTON    83   42  101   39   174  190   82   105   45
SIMON          38   33   54   37    82   82  111    58   37
SPRAGUE        67   59  133   49   126  105   58   182   61
WHINSTON       27   93   54   24    50   45   37    61  104
;
```
%INCLUDE 'c:\program files\SAS Institute\SAS\V8\stat\sample\xmacro.sas';
%INCLUDE 'c:\program files\SAS Institute\SAS\V8\stat\sample\distnew.sas';
%DISTANCE (data=aca9var,
    id=author,
    options=print,
    method=dcorr,
    var=x1-x9,
    out=dist);
RUN;

PROC CLUSTER METHOD=WARD DATA=dist PSEUDO;
ID author;
PROC TREE SPACE=**2**;

RUN;

INTERPRETING RESULTS OF CLUSTER ANALYSIS

A distance matrix derived from a transposed cocitation frequency matrix is the input to the PROC CLUSTER procedure of the SAS system (release 8.01). The CLUSTER procedure with Ward's trace option produces data set WORK.DIST, Custer History, and Proc Tree Graph Output.

Table 5.6 Data Set Work.DIST

Obs	Author	ALTER	BONCZEK	CARLSON	HUBER
1	ALTER	0.00000	.	.	.
2	BONCZEK	1.21223	0.00000	.	.
3	CARLSON	0.74515	1.01124	0.00000	.
4	HUBER	1.07661	1.23667	0.98788	0.00000
5	KEEN	0.56310	1.14216	0.61729	0.89764
6	SCOTTMORTON	0.58664	1.14668	0.66094	0.92065
7	SIMON	0.91242	1.17918	0.89841	0.91234
8	SPRAGUE	0.76448	1.00294	0.35056	0.88388
9	WHINSTON	1.20648	0.13646	1.00781	1.23657

Obs	KEEN	SCOTTMORTON	SIMON	SPRAGUE	WHINSTON
1	.	.	.	.	.
2	.	.	.	.	.
3	.	.	.	.	.
4	.	.	.	.	.
5	0.00000	.	.	.	.
6	0.16624	0.00000	.	.	.
7	0.66143	0.62572	0.00000	.	.
8	0.58349	0.65439	0.83201	0.00000	.
9	1.12845	1.13344	1.16098	0.99648	0

DATA SET WORK.DIST

The first part of the output (WORK.DIST) is a triangle shape of distance matrix. Any dataset begins with WORK is a temporary SAS data set. Temporary data sets are deleted at the end of the session. As shown in Table 5.6, the first column lists variables and the first row lists distance variables. The intersection of observation 1 and distance 1 contains computed relative distance ("0") between the first author itself. The intersecting cell of observation 2 and dist 1 column contains ".71708", a relative distance between ALTER and ACKOFF.

CLUSTER HISTORY

The next output shows cluster history (Table 5.7).

Table 5.7 Cluster History

NCL	------Clusters Joined------		FREQ	SPRSQ	RSQ	PSF	PST2	Tie
8	BONCZEK	WHINSTON	2	0.0028	.997	51.3	.	
7	KEEN	SCOTTMORTON	2	0.0041	.993	48.0	.	
6	CARLSON	SPRAGUE	2	0.0183	.975	23.2	.	
5	ALTER	CL7	3	0.0644	.910	10.2	15.6	
4	CL5	SIMON	4	0.1068	.804	6.8	3.1	
3	CL4	CL6	6	0.1446	.659	5.8	3.0	
2	CL3	HUBER	7	0.1825	.477	6.4	2.7	
1	CL2	CL8	9	0.4766	.000	.	6.4	

With the PRINT= option, The PROC CLUSTER output displays the following.

Column 1 (NCL -- the Number of Clusters): With N number of variable dataset, the maximum number of cluster is N-1. Table 5.7 shows the maximum number of cluster 8.

Columns 2 and 3 (the names of the Clusters Joined): The observations are identified by the formatted value of the ID variable, if any; otherwise, the observations are identified by OBn, where n is the observation number. Clusters of two or more observations are identified as CLn, where n is the number of clusters existing after the cluster in question is formed.

Column 4 (FREQ is the number of observations in the new cluster). The first row's FRQ is 2. Cluster 8 is formed by joining 2 observations (Bonczek and Whinston). The last row, cluster 1, has all 9 authors in it.

Column 5 (SPRSQ stands for Semipartial R-Squared) SPRSQ indicates the decrease in the proportion of variance accounted for resulting from joining the two clusters. This equals the between-cluster sum of squares divided by the corrected total sum of squares.

Column 6 (RSQ is the squared multiple correlation, R-Squared). R^2 is the proportion of variance accounted for by the clusters. For example, all 8 clusters account for 99.7% of variance in the dataset.

Column 7 (PSF is Pseudo F). The pseudo F statistic measures the separation among all the clusters at the current level.

Column 8 (PST2 is Pseudo t^2). The pseudo t^2 statistic measures the separation between the two clusters most recently joined.

For more detailed information on this subject, readers are referred to (Khattree et al. 1999; SAS Institute Inc. 2000). Users at academic institutions can access SAS OnlineDoc®, Version 8, HTML format with no charge after registration at http://v8doc.sas.com/sashtml/

SELECTING THE NUMBER OF CLUSTERS

The proc cluster procedure produces cluster history beginning with the number of cluster 1 to N-1 with the N number of variable dataset. Table 5.7 exhibits the number of clusters ranging 8 through 1, since the data set has 9 authors. One of the important decisions for ACA analysts is the selection of the optimal number of clusters. There are no structured procedures in determining the number of clusters. The optimal number of clusters can be chosen by using one or more of the following: (1) the cubic clustering criterion (CCC), (2) the pseudo F statistic, (3) the pseudo t^2 statistic. The cubic clustering criterion (CCC) can only be applied if the data are coordinates. Since the ACA uses distance data,) the pseudo F statistic and the pseudo t^2 statistic are most frequently used.

PLOT OF THE PSEUDO F STATISTIC AGAINST NUMBER OF CLUSTERS

Plots of the pseudo F statistic against the number of clusters are useful to select the number of clusters. Figure 5.2 shows the plot. The peaks of pseudo F values in the plot can be used to decide the number of clusters. Depending on the number of variables, ACA analysts may ignore the pseudo F values of the last 2 or 3 clusters (e.g., clusters 7 and 8). Figure 5.2 has peaks at 2, 4, 5, and 6 clusters. Due to small number of variables used to produce the plot, this example may not be good for the beginners in ACA. Figure 5.3 is based on the data set used in SAS/STAT® User's Guide (Release 6.03 Edition). In this example, there are three peaks at clusters 3, 8, and 12, ignoring the last two clusters (14 and 15).

Figure 5.2 Plot of Pseudo F Statistic against Number of Clusters

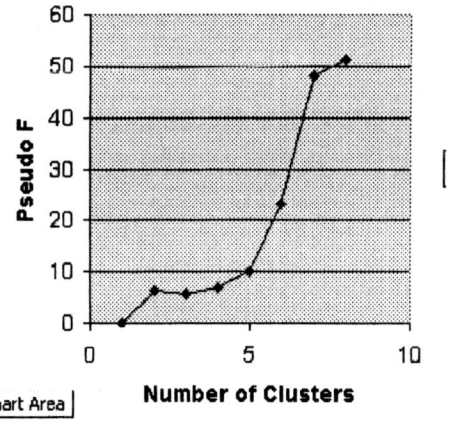

Figure 5.3 Plot of Pseudo F Statistic against Number of Clusters using Birth and Death Rates in 74 Countries

1	2	3	4	5	6	7	8	9	10	11	12	13	14	15
0	271.61	279.3	253.49	234.97	243	261.76	296.95	295.08	296.24	304.44	322	320.41	324.11	325.9

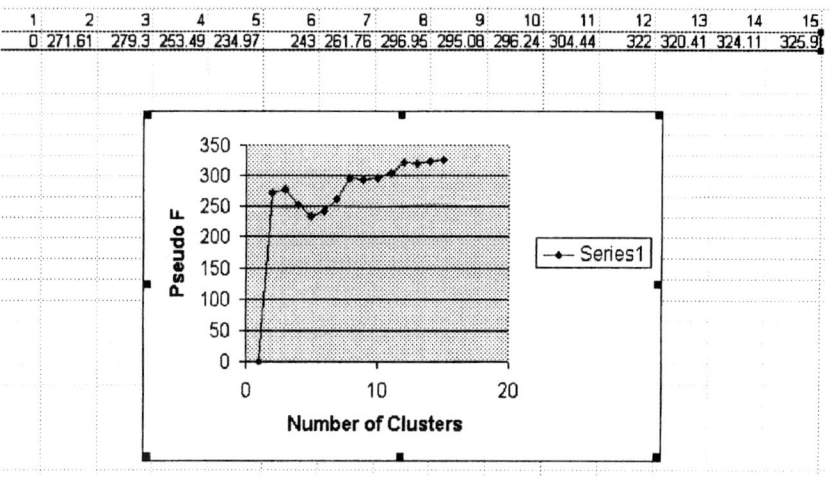

PLOT OF THE PSEUDO t^2 STATISTIC AGAINST NUMBER OF CLUSTERS

Plots of the pseudo t^2 statistic against the number of clusters are also useful to select the number of clusters (Figures 5.4 and 5.5). The valleys of pseudo F values in the plot can be used to decide the number of clusters. Depending on the number of variables, ACA analysts may ignore the values of the last 2 or 3 clusters. Figure 5.4 has valleys at 2, 3, and 4 clusters. In Figure 5.5, the pseudo t^2 statistic indicates 3, 8, and 12 clusters.

Considering the pseudo t^2 statistic and the pseudo F statistic simultaneously narrows down the number of clusters for final selection. In the case of the 9 variable ACA example we have used, the 2 and 4 cluster solutions are suggested for further evaluation. The cluster analysis of the birth and death rates data in 74 countries suggests 3, 8, and 12 cluster solution for further evaluation. The interpretability of cluster solution is the decisive criteria in selecting the number of cluster. ACA analysts should be able to name each cluster based on contents of each author under each cluster.

Figure 5.4 Plot of Pseudo t^2 Statistic against Number of Clusters

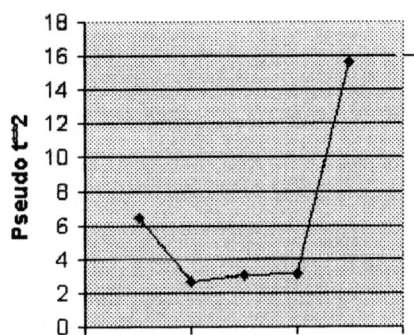

Figure 5.5 Plot of Pseudo t^2 Statistic against Number of Clusters using Birth and Death Rates in 74 Countries

1	2	3	4	5	6	7	8	9	10	11	12	13	14	15
271.61	69.12	30.71	41.03	17.07	20.37	16.09	11.92	16.33	11.54	11.48	8.8	12.32	13.13	5.12

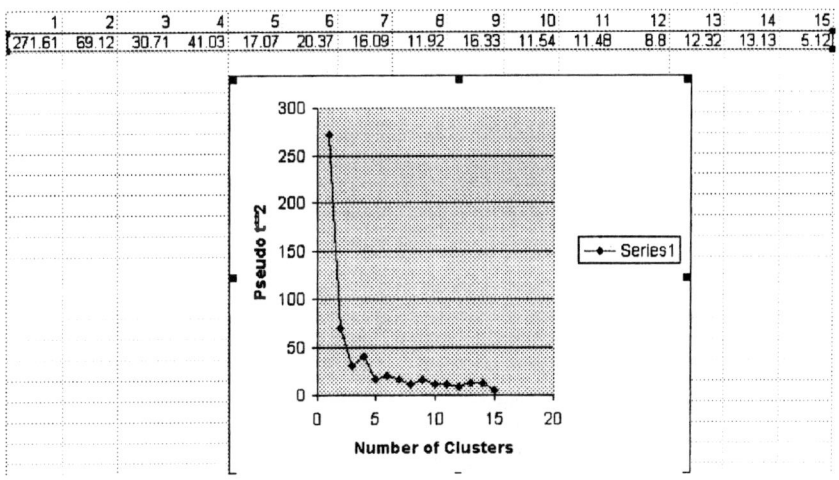

PROC TREE GRAPH OUTPUT

The final output is the Tree graph (dendrogram). The Dendrogram illustrates hierarchical clustering of eight authors of decision support systems researchers. Figure 5.6 shows both the cluster structure and the joining sequence to show how each of the authors in the study is combined into a new aggregate cluster until all 9 authors are grouped into the final one cluster, cluster one (CL1).

Figure 5.6 Dendrogram (Tree Graph) Depicting Cluster Structure and Joining Sequences

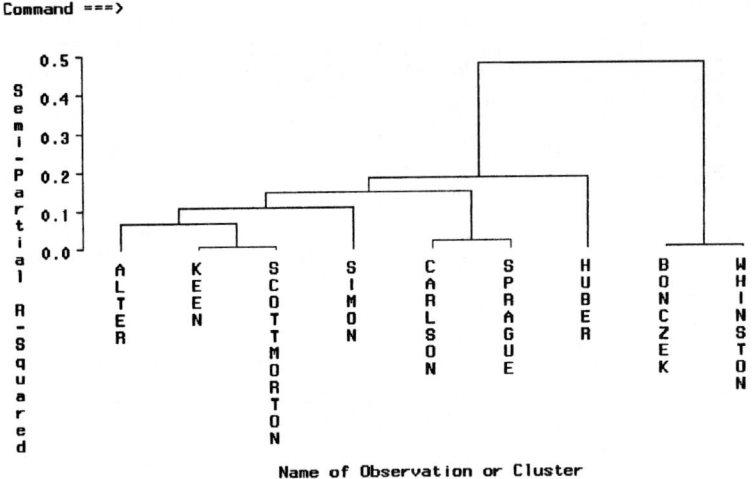

CL 8 = (Bonczek, Whinston)

CL 7 = (Keen, Scott Morton)

CL 6 = (Carlson, Sprague)

CL 5 = (Alter, Keen, Scott Morton)

CL 4 = (Alter, Keen, Scott Morton, Simon)

CL 3 = (Alter, Keen, Scott Morton, Simon, Carlson, Sprague)

CL 2 = (Alter, Keen, Scott Morton, Simon, Carlson, Sprague, Huber)

CL 1 = (Bonczek, Whinston, Alter, Keen, Scott Morton, Simon, Carlson, Sprague, Huber)

In Table 5.6 which displays the distance among each author, the shortest distance between two authors is .13646 between Bonczek and Whinston, followed by .16624 between Keen and Scott Morton, and .35056 between Sprague and Carlson. Based on the proximity of these author set, clusters 8, 7, and 6 are formed successively as shown in Table 5.7. In the next subsequent step, cluster 3 is combined with cluster 5 to form a new cluster (CL2). The dendrogram shows a linkage among various clusters, and influence could be inferred based on the close examination of the works of authors in the clusters. In the final agglomerative clustering procedure, Cluster 8 (Bonczek and Whinston) and cluster 2 (cluster of all other authors except two authors in Cluster 8) are the last clusters to be joined to form the final cluster (CL1), which indicates the heterogeneity of the authors of cluster 2.

Comparison of the two solutions from the factor analysis (Table 4.15) and cluster analysis (Figure 5.6) provides some valuable information on the similarities and differences of the two solutions to help us reach a better interpretation of the results of multivariate analysis. In this example of small variables, factor solution and cluster solution coincide very well. Cluster 3 (6 authors) corresponds to factor 1. Factor 2 coincides with cluster 8. Factor 3 is a factor with only one author (Huber). In ACA, single factor solution is not generally acceptable. Statistically, it may be an acceptable solution. But in the study of author cocitation, we are interested in the linkage among authors. Therefore, if the single author factor appears in the final solution in any ACA study, ACA researchers do not accept the factor solution. In that case, the tree graph from cluster analysis can provide a useful guide when assigning a single author to any other factors. See several previous studies of Eom (1997; 1998b) to see how factor analysis and cluster analysis can be complementary each other in ACA.

CHAPTER 6 MUTIDIMENSIONAL SCALING

This chapter discusses multidimensional scaling (MDS) procedures. MDS is a class of multivariate statistical techniques/procedures to produce two or three dimensional pictures of data (geometric configuration of points) using proximities among any kind of objects as input. The purposes of MDS are to help researchers identify the "hidden structures" in the data and visualize relationships among/within the hidden structures to give clearer explanations of these relationships to others (Hair et al. 1987; Kruskal et al. 1990). Three SAS procedures (MDS, PLOT, and G3D) are necessary to convert the author cocitation frequency matrix to two or three dimensional pictures of data.

The distance matrix produced earlier by using xmacro.sas and distnew.sas programs should be converted to a coordinate matrix. The coordinate matrix is used to produce two-dimensional plots and annotated three-dimensional scatter diagrams. A distance matrix is the input to the multidimensional scaling procedure, PROC MDS, of the SAS system (release 8.0). The PLOT and G3D procedures process the coordinate matrix to visualize the similarity and dissimilarity within each group of an academic discipline as well as the similarity and dissimilarity among the various subspecialties within an academic discipline. In ACA study, 3D scatter plots without labels on data points provide little information for the ACA researchers. This chapter also discusses how to label data points on a plot. The annotate facility in the SAS system produces figures with the name of the author on each data point. The PROC MDS procedure includes many of the features of the ALSCAL procedure (Young et al. 1986) and some features of the MLSCALE procedure (Ramsay 1986) (SAS Institute Inc. 1992).

THE MDS PROCEDURE

Multidimensional scaling is a multivariate statistical analysis tool for examining proximity data among any kind of object. Proximity data consist of one or more square symmetric or asymmetric matrices of similarities or dissimilarities between *objects* or *stimuli* (Kruskal et al. 1978, pp. 7-11). The MDS outputs consist of a spatial representation of data which shows underlying relationships on a two or three dimensional map. The MDS map helps visualize relationships more clearly using the ratio of distances on a map to corresponding data values such as a map of a country showing cities. The magnitude of number indicates how similar/dissimilar two objects are.

SIMILARITY/PROXIMITY MEASURES

How should the inter-object similarity be measured? Numerous ways of measuring inter-object similarity exist. The non-metric data measures the distance by directly ranking the objects from most preferred to least preferred (preference data) and using the pairwise comparison (similarity data) to determine which items are most similar/dissimilar to each other (all pairs of these objects can be compared).

To measure proximities among authors, the correlations among authors are used most frequently. Correlations are used as proximities by MDS procedures

(Kruskal et al. 1978) The author cocitation frequency is metric data. As in the PROC Cluster procedure, the cocitation frequency matrix must be converted into a ordinary Euclidian distance data matrix using METHOD= DCORR, This method transformed correlations to Euclidean distance using square root of (1-CORR). Tables 6.1 and 6.8 show MDS SAS procedure statements.

Figure 6.1 MDS/PLOT/G3D Procedures in ACA

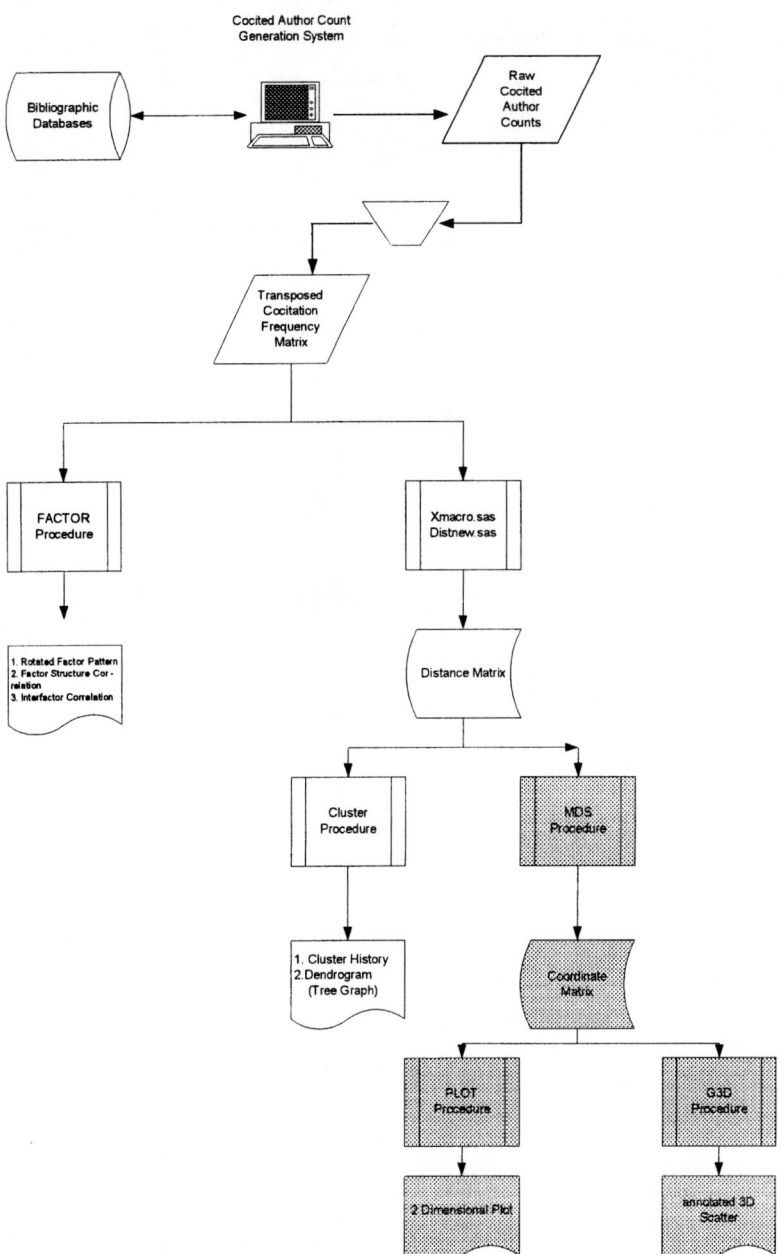

Table 6.1 PROC MDS with a Permanent Data Set (Mysave.dist)

libname mysave 'c:\wp\books\aca\sasdata\perm';
PROC MDS
    data=mysave.dist
    condition=un
    level=ratio
    coef=identity
    dimension=**3**
    cutoff=**0**
    fit=**2**
    out=mysave.coord
    outres=res;
    ID author;
    pcoef
    pconfig
    pfinal
;
RUN;

PROC PRINT data=mysave.coord (keep=author dim1 dim2 dim3);
VAR author dim1 dim2 dim3;
RUN;
options ps=**40**;

PROC MDS STATEMENT

The MDS procedure produces only the iteration history and it estimates the coordinate of a set of objects in a space of specified dimensionality. The procedure does not produce any graphical outputs (SAS Institute Inc. 2000).

PROC MDS < options > ;
    VAR variables ;
    INVAR variables ;
    ID | OBJECT variable ;
    MATRIX | SUBJECT variable ;
    WEIGHT variables ;
    BY variables ;

Since the MDS procedure produces only the printed iteration history, specifying options are necessary to produce other results listed below.

SPECIFY THE INPUT DATA SET

>DATA=*SAS-data-set*

SPECIFY THE OUT DATA SET

>OUT=*SAS-data-set*

>This option creates a SAS data set containing the estimates of all the parameters of the MDS model and the values of the badness-of-fit criterion.

>OUTRES=SAS-data-set

>This option creates a SAS data set with one observation for each non-missing datum from the DATA=data set. Each observation contains the original datum, the estimated distance computed from the MDS model, transformed data and distances, and the residual.

SPECIFY THE TYPE OF THE DIMENSION COEFFICIENTS

>COEF= {Identity, Diagonal}
>This option specifies the method of computing the dimension coefficients of each variable (author). The coefficients can be Euclidean distances (Identity) or weighted Euclidean distances, in which each subject is allowed differential weights for the dimensions.

SPECIFY THE CONDITIONALITY OF THE DATA

> CONDITION=UN
> This option puts all the data into a single partition. Each data matrix is stored as a triangle.

SPECIFY THE MEASUREMENT LEVEL OF THE DATA AND THE TYPE OF TRANSFORMATIONS

> LEVEL={absolute, ratio, interval, ordinal, loginterval, ordinal}
>
> LEVEL=RATIO option specifies a linear transformation that fits a regression model in which the distances are multiplied by a slope parameter in each partition (a linear transformation). In this case, the regression model is equivalent to the measurement model with the slope parameter reciprocated.

SPECIFY THE NUMBER OF DIMENSIONS

> DIMENSION = n
>
> This option specifies the number of dimensions to use in the MDS model.

SPECIFY A PREDETERMINED TRANSFORMATION

> FIT = {Distance, Squared, Log, N}
>
> This option specifies a predetermined (not estimated) transformation to apply to both sides of the MDS model before the error term is added. FIT=SQARED or FIT=2 fits squared data to squared distances.

MISCELLANEOUS

CUTOFF = n

This option specifies that data less than n is to be replaced by missing values. The default value is CUTOFF =0.

ID STATEMENT

The ID statement specifies a variable in the DATA= data set that contains descriptive labels (author's name) for each author. The name of authors are used in the output and are copied to the OUT= data set, instead of non-descriptive variable names such as var1.

CONTROL DISPLAYED OUTPUT

According to the SAS/STAT user's guide of Online SAS Manual Version 8, displayed output is the results of many interacting options.

Displayed output is controlled by the interaction of the PCONFIG, PCOEF, PTRANS, PFIT, and PFITROW options with the PININ, PINIT, PITER, and PFINAL options. The PCONFIG, PCOEF, PTRANS, PFIT, and PFITROW options specify *which* estimates and fit statistics are to be displayed. The PININ, PINIT, PITER, and PFINAL options specify *when* the estimates and fit statistics are to be displayed. If you specify at least one of the PCONFIG, PCOEF, PTRANS, PFIT and PFITROW options but none of the PININ, PINIT, PITER, and PFINAL options, the final results (PFINAL) are displayed. If you specify at least one of the PININ, PINIT, PITER, and PFINAL options but none of the PCONFIG, PCOEF, PTRANS, PFIT and PFITROW options, all estimates (PCONFIG, PCOEF, PTRANS) and the fit statistics for each matrix and for the entire sample (PFIT) are displayed. If you do not specify any of these nine options, no estimates or fit statistics are displayed (except the badness-of-fit criterion in the iteration history).

PCOEF prints the estimated dimension coefficients. The dimension coefficient is the square roots of the *subject weights* (Kruskal et al. 1978). The dimension coefficient for each data matrix is "the coefficients that multiply each coordinate of the *common* or *group* weighted Euclidean space to yield the *individual* unweighted Euclidean space" (SAS Institute Inc. 2000).

PCONFIG prints the estimated coordinates of the objects in the configuration.

PFINAL prints final estimates.

Table 6.2 MDS Procedure Output

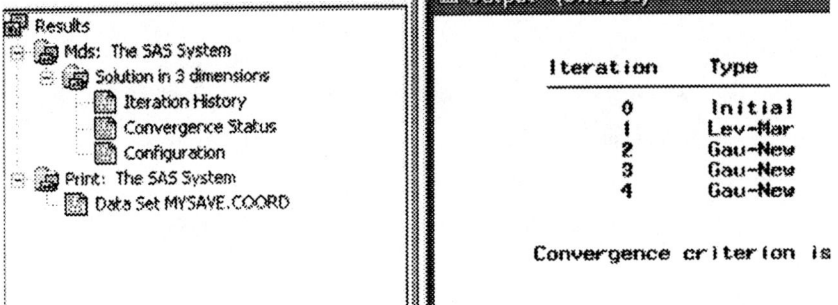

MDS PROCEDURE OUTPUT

Table 6.2 shows four outputs from the PROC MDS procedure (Iteration history, Convergence Status, Configuration, and Data Set MYSAVE.COORD).

<u>ITERATION HISTORY AND CONVERGENCE STATUS</u>

The current solution is obtained after 4 iterations. PROC MDS displays the iteration history containing

- Iteration number (0 through 4)

- Type of iteration: Initial (initial configuration), Monotone (monotone transformation), Gau-New (Gauss-Newton step), and Lev-Mar (Levenberg-Marquardt step).

- Badness-of-Fit Criterion: Using formula=1, The badness-of-fit criterion is square root of $1-R^2$, where R^2 is the percent of explained variability.

The badness of fit represents square root of unexplained variability. Fit can be measured by the square root of multiple correlation (R^2) or square root of $1-R^2$. The former is often called goodness-of-fit and the latter is called badness-of-fit. Good fit values are near zero and poor values are near one, if we use badness-of-fit criterion.

- Changes in criterion refers to the difference of badness-of-fit criterion between previous alteration and current alteration. For example, the value, .0207, is the difference between the badness-of-fit value (.1365) at iteration 0 and (.1158) that of iteration 1.

Table 6.3 Iteration History and Convergence Status

```
                         MDS output

             Multidimensional Scaling:  Data=WORK.DIST
             Shape=TRIANGLE Condition=UN Level=RATIO
             Coef=IDENTITY Dimension=3 Formula=1 Fit=2

        Gconverge=0.01 Maxiter=100 Over=1 Ridge=0.0001


                         Badness-
                          of-Fit      Change in     Convergence
     Iteration    Type   Criterion    Criterion      Measure
          0     Initial    0.1365          .           0.4907
          1     Lev-Mar    0.1158       0.0207         0.1110
          2     Gau-New    0.1148       0.000944       0.0375
          3     Gau-New    0.1147       0.000113       0.0158
          4     Gau-New    0.1147       0.0000214      0.008107

        Convergence criterion is satisfied.
```

Table 6.4 Configuration

Configuration

	Dim1	Dim2	Dim3
ALTER	1.05	1.25	-0.34
BONCZEK	-2.32	0.07	-0.27
CARLSON	0.17	0.71	1.04
HUBER	0.48	-1.93	0.89
KEEN	1.01	0.26	-0.30
SCOTTMORTON	1.00	0.22	-0.58
SIMON	0.71	-1.00	-1.26
SPRAGUE	0.18	0.35	1.13
WHINSTON	-2.28	0.07	-0.32

Table 6.4 shows a configuration of all authors on a three-dimensional space. The configuration of points is the coordinates of each object in a Euclidean space of one or more dimensions (Kruskal et al. 1978; Young et al. 1987). The distances between data points are constant over any similarity transformation such as rotation, permutation, reflection, translation, and dilation of the dimensions.

Table 6.5 Data Set MYSAVE.COORD

MDS output

Obs	Author	Dim1	Dim2	Dim3
1		0.11468	.	.
2	ALTER	1.04522	1.24929	-0.34354
3	BONCZEK	-2.31542	0.07483	-0.26530
4	CARLSON	0.17294	0.71272	1.04272
5	HUBER	0.47546	-1.93306	0.88987
6	KEEN	1.01034	0.25904	-0.30261
7	SCOTTMORTON	1.00292	0.22087	-0.58030
8	SIMON	0.71084	-1.00424	-1.25772
9	SPRAGUE	0.17918	0.34683	1.13255
10	WHINSTON	-2.28148	0.07373	-0.31568
11		0.34472	.	.

THE PLOT PROCEDURE

As mentioned earlier, the MDS procedure produces only the printed iteration history and coordinate matrix. PROC MDS does not produce any plots. Additional plot procedures are necessary to produce two or three dimensional graphs. This section describes the production of two dimensional scatter plots to show the relationships between two or more dimensions. There are many different types of two dimensional plots such as line, high-low, bubble, and scatter. Scatter plots best represent the coordinate matrix. Figure 6.1 shows that the input to the plot procedure is the coordinate matrix produced by the MDS procedure and saved as a permanent data set, mysave.coord. The PLOT procedure plots the coordinates of each author in a Euclidean space of one or more dimensions.

PROC PLOT STATEMENT

The basic syntax of producing a two dimensional scatter plot consists of PROC PLOT statement with options, followed by another PLOT statement with options. The following example shows the basic syntax to produce Figure 6.2.

```
libname mysave 'c:\wp\books\aca\sasdata\perm';
PROC PLOT data=mysave.coord;
   PLOT dim1*dim2 = '*' $ author;
RUN;
```

The PROC PLOT statement uses an option, data=data-set name. The syntax of PLOT statement is: **PLOT** < *yvariable**xvariable >< =*symbol* >. The statement, PLOT dim1*dim2 = '*' $ author, tells the SAS system that dimension 1 variables and dimension 2 variables in Table 6.5 be plotted on the vertical (y coordinate) and horizontal (x coordinate) axes respectively. It also tells that the symbol with single quotation (*) will be used to represent each variable on the plot and character label of author name ($) will be on the plot.

Figure 6.2 Plot of Dim1* Dim2$Author

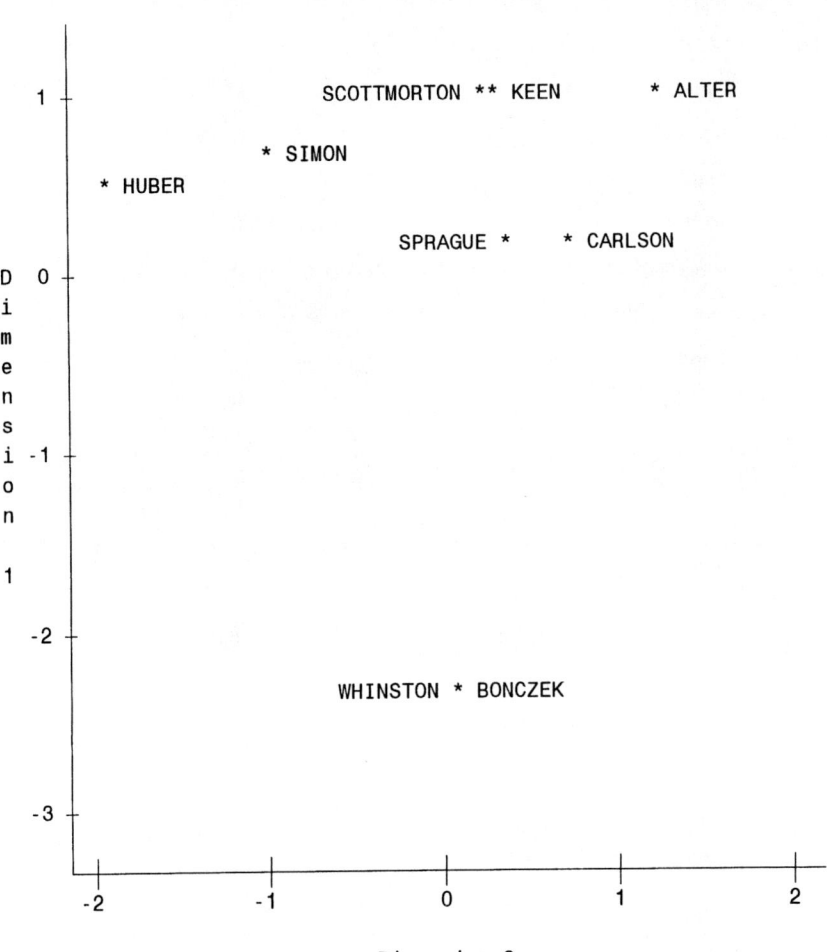

PROC PLOT <*option(s)*>;

    PLOT *plot-request(s)* </ *option(s)*>;

        Plot-request(s) specifies the vertical and horizontal variables to plot and the plotting symbol to use to mark the points on the plot.

PROC PLOT < OPTION(S)>

<u>SPECIFY THE INPUT DATA SET</u>

        DATA= SAS-data-set

<u>CONTROL THE APPEARANCE OF THE PLOT</u>

        VTOH=*aspect-ratio*

        VTOH stands for Vertical **TO** **H**orizontal. This option specifies the aspect ratio (vertical to horizontal) of the plot. *Aspect-ratio* is a positive real number.

Table 6.6 PLOT *plot-request(s) </ option(s)>*;

	Use this option
Control the axes	
Specify the tick-mark values	HAXIS=, VAXIS=
Expand the axis	HEXPAND, VEXPAND
Specify the number of print positions	HPOS=, VPOS=
Reverse the order of the values	HREVERSE, VREVERSE
Specify the number of print positions between tick marks	HSPACE=, VSPACE=
Assign a value of zero to the first tick mark	HZERO and VZERO
Specify reference lines	
Draw a line perpendicular to the specified values on the axis	HREF= and VREF=
Specify a character to use to draw the reference line	HREFCHAR= and VREFCHAR=
Put a box around the plot	BOX
Produce a contour plot	
Draw a contour plot	CONTOUR
Specify the plotting symbol for one contour level	S*contour-level*=
Specify the plotting symbol for multiple contour levels	SLIST=
Label points on a plot	
List the penalty and the placement state of the points	LIST=
Force the labels away from the origin	OUTWARD=
Change default penalties	PENALTIES=
Specify locations for the placement of the labels	PLACEMENT=
Specify a split character for the label	SPLIT=

PLOT PLOT-REQUEST(S) </ OPTION(S)>

The PLOT statement followed by the PROC PLOT statement requests the plots to be produced by PROC PLOT. Most of The PLOT statement options are listed in Table 6.6, taken from (SAS Institute Inc. 2000). The following options are used in the example used in ACA plot procedures.

<u>CONTROL AXES BY SPECIFYING THE TICK-MARK VALUES</u>

HAXIS = specifies major tick mark values for horizontal axis of box plot.

VAXIS = specifies major tick mark values for vertical axis of box plot.

The AXIS Statement is to control the location, values, and appearance of the axes of the plots and charts. This statement specifies the characteristics of an axis such as:

(1) the way the axis is scaled
(2) how the data values are ordered
(3) the location and appearance of the axis an the tick mark
(4) the text and appearance of the axis label and major tick mark values.

TYPE= 'CONFIG' specifies that the output of this PLOT procedure is a spatial arrangement on a two/three dimensional space.

<u>USE A BOX AROUND THE PLOT</u>

The BOX option is used to draw a box around the plot. Without the box option, the plot procedure produces a plot without a box. See Figure 6.2.

Example: PLOT (dim1 dim2) * dim3 = '*' $ author / BOX

Figure 6.3 PROC PLOT Statement with the HAXIS, YAXIS Options

LIBNAME mysave 'c:\wp\books\aca\sasdata\perm';
PROC PLOT data=mysave.coord;
  PLOT dim1*dim2 = '*' $ author
   / HAXIS=BY .5
    VAXIS=by .5;
    where _type_='CONFIG';
  PLOT (dim1 dim2) * dim3 = '*' $ author
   / HAXIS=BY .5
    VAXIS=by .5;
    where _type_= 'CONFIG';
RUN;

Figure 6.4 PROC PLOT Statement with the BOX, VTOH Options

LIBNAME mysave 'c:\wp\books\aca\sasdata\perm';
PROC PLOT data=mysave.coord vtoh=1;
  PLOT dim1*dim2 = '*' $ author
   / BOX
    HAXIS=BY .5
    VAXIS=by .5;
    where _type_='CONFIG';
  PLOT (dim1 dim2) * dim3 = '*' $ author
   / BOX
    HAXIS=BY .5
    VAXIS=by .5;
    where _type_= 'CONFIG';
RUN;

Figure 6.5 Plot of Dim1*Dim2$Author with HAXIS, YAXIS Options

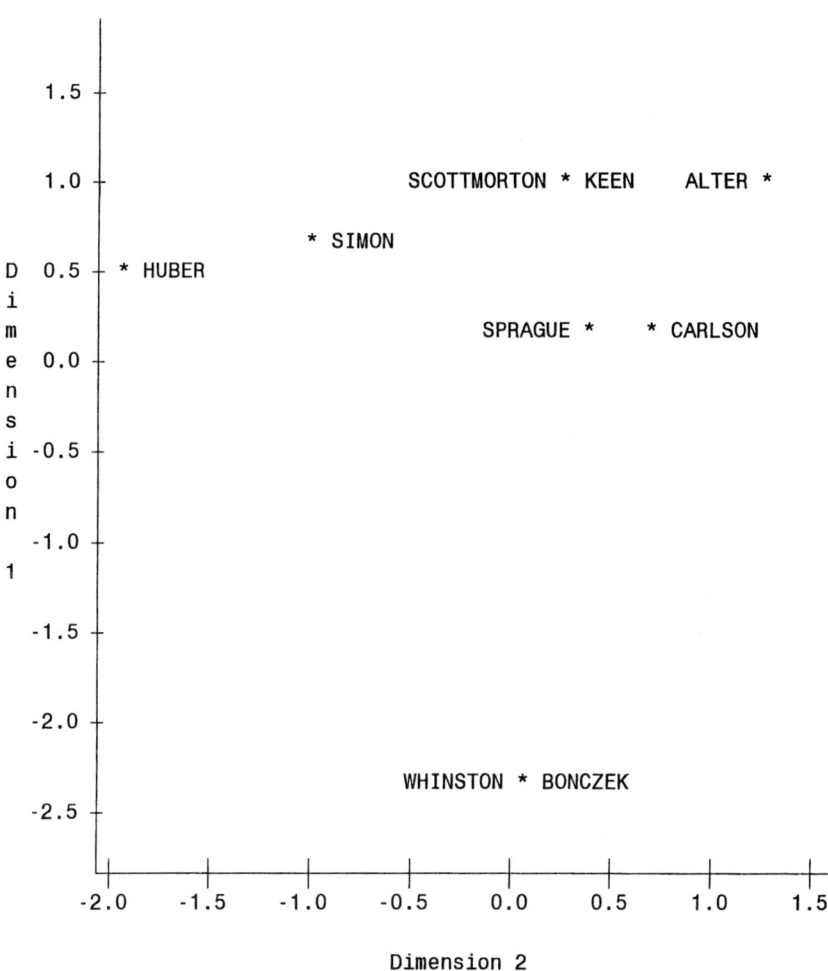

NOTE: 2 obs hidden.

Figure 6.6 Plot of Dim1*Dim2$Author with VTOH = 1

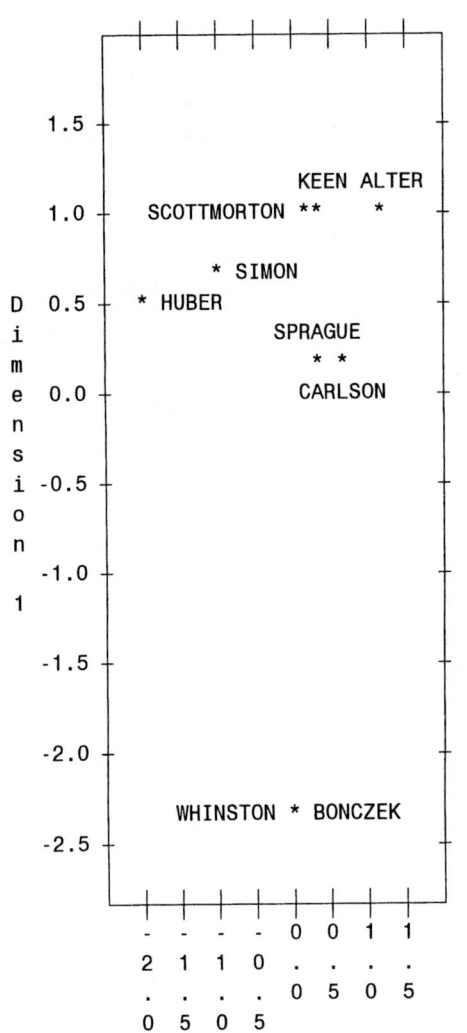

NOTE: 1 obs hidden.

Figure 6.7 Plot of Dim1*Dim2$Author with VTOH=3

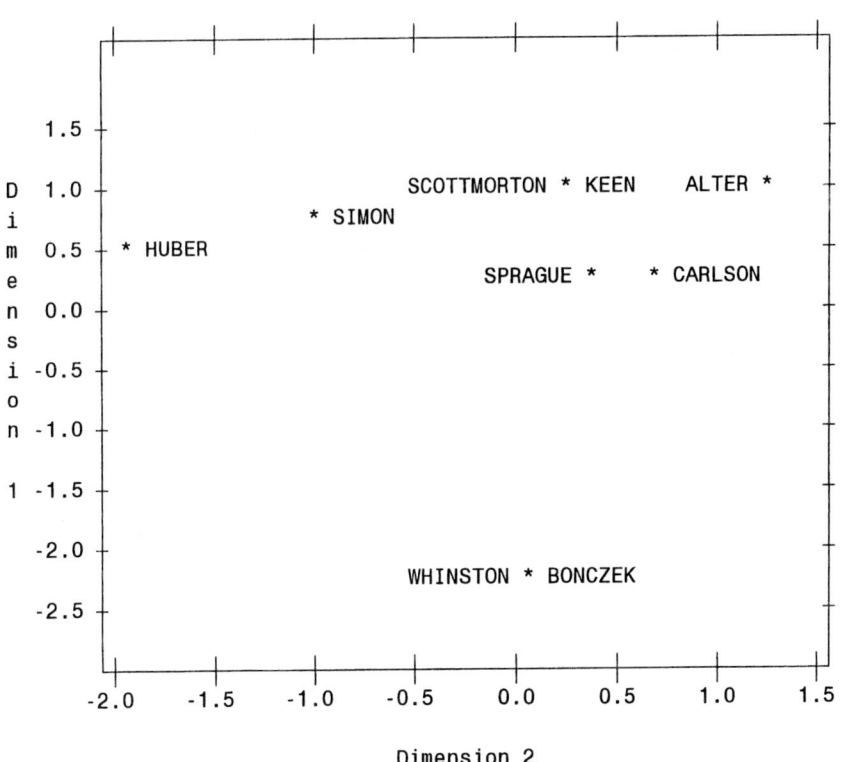

NOTE: 2 obs hidden.

PLOT PROCEDURE OUTPUT

Figures 6.8 and 6.9 present two dimensional MDS maps projecting the first and third planes (Figure 6.8) and the second and third planes (Figure 6.9) presented in the three dimensional maps in Figure 6.12. While cluster analysis shows the detailed structure of clusters, multidimensional scaling maps show the big picture of inter-cluster relationships. The placement of authors on the center of the MDS map means that those authors are linked with a substantial portion of the author set, with relatively high correlations. The center of the MDS maps has the value of zero on the horizontal axis as well as on the vertical axis. Placement near the periphery represents a more focused linkage.

Figure 6.8 Plot of Dim1*Dim3

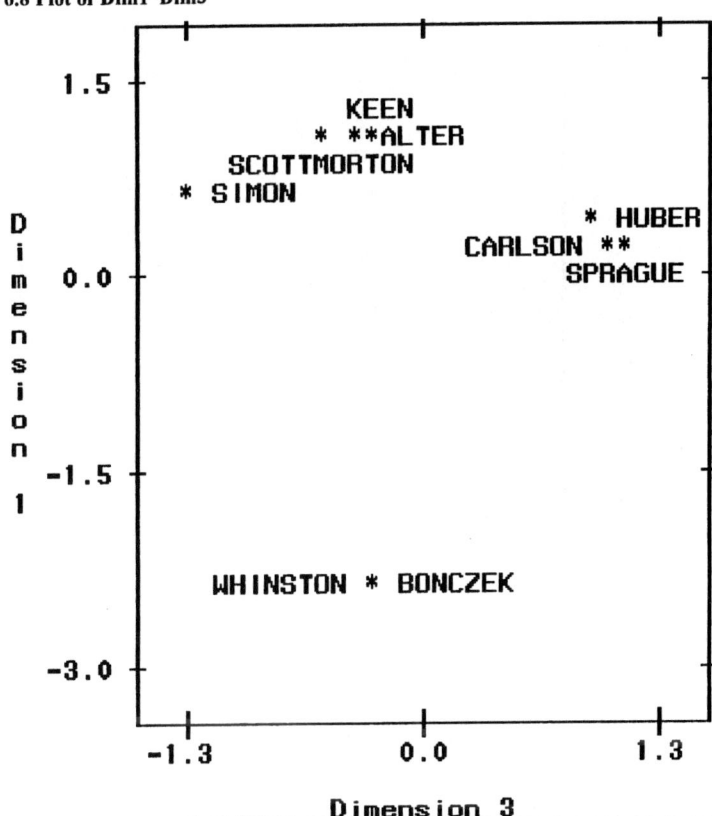

Figure 6.9 Plot of Dim2*Dim3

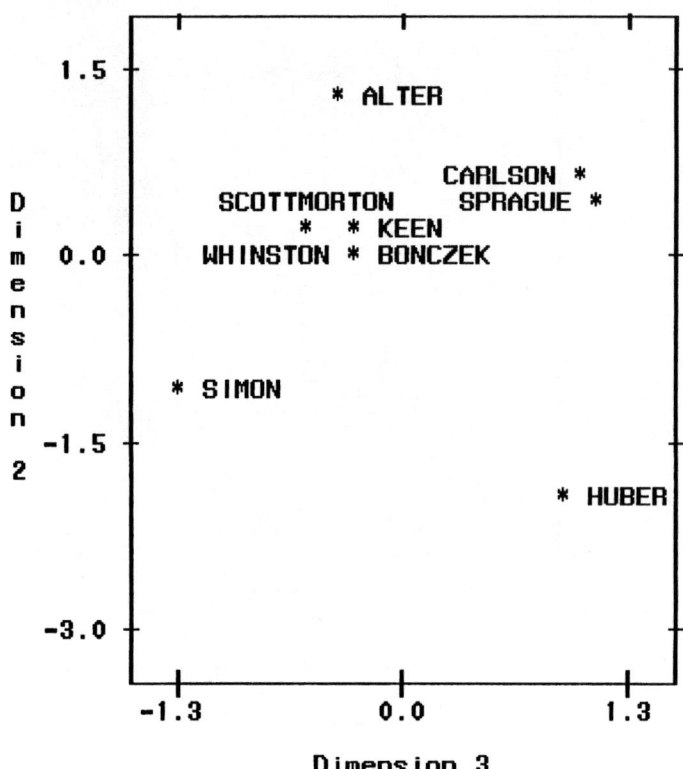

Table 6.7 PROC MDS SAS STATEMENT

```
DATA test1;
INPUT Author $ 1-12 @13 x1-x9;
CARDS;
ALTER        121  25  66  23   92   83  38  67   27
BONCZEK       25 103  53  23   46   42  33  59   93
CARLSON       66  53 173  34  112  101  54 133   54
HUBER         23  23  34  68   47   39  37  49   24
KEEN          92  46 112  47  206  174  82 126   50
SCOTTMORTON   83  42 101  39  174  190  82 105   45
SIMON         38  33  54  37   82   82 111  58   37
SPRAGUE       67  59 133  49  126  105  58 182   61
WHINSTON      27  93  54  24   50   45  37  61  104
;
%include 'c:\program files\SAS Institute\SAS\V8\stat\sample\xmacro.sas';
%include 'c:\program files\SAS Institute\SAS\V8\stat\sample\distnew.sas';
%distance ( data=test1,
            id=author,
            options=print,
            method=dcorr,
            var=x1-x9,
            out=dist);
proc print data=dist;
run;

/**** MDS procedures to produce profiles****/
proc mds
      data=dist
      condition=un
      level=ratio
      coef=identity
      dimension=3
      cutoff=0
      fit=2
      out=coord
      outres=res;
      pcoef
      pconfig
      pfinal
      ID author;
;
title1 'MDS Output';
run;
proc print
```

```
data=coord (keep=author dim1 dim2 dim3);
var author dim1 dim2 dim3;
run;
options ps=40;

proc plot
      data=coord vtoh=.5;
      plot dim1*dim2 = '*' $ author / BOX
      HAXIS=BY .4 VAXIS=by 1.3;
      where _type_='CONFIG';
      plot (dim1 dim2) * dim3 = '*' $ author / BOX
      HAXIS=BY .35 VAXIS=by .35;
      where _type_ = 'CONFIG';

      TITLE2 'PLOT Of CONFIGURATION';
data anno;
set coord (keep=dim1 dim2 dim3 author);
length function $8;
retain xsys ysys zsys '2'
function 'label'
      size .4
      position '2'
      color 'blue'
      style 'zapf';
      x=dim1;
      y=dim2;
      z=dim3;
      text=author;

proc g3d
      data=coord;
      scatter dim2*dim1=dim3 /anno=anno shape='pyramid';
run;
quit;
```

Table 6.8 PROC MDS with a Permanent Data Set (Mysave.dist)

libname mysave 'c:\wp\books\aca\sasdata\perm';
proc mds
    data=mysave.dist
   condition=un
   level=ratio
   coef=identity
   dimension=**3**
   cutoff=**0**
   fit=**2**
   pcoef
   pconfig
   pfinal
   out=mysave.coord
   outres=res;
   ID author;
;
title1 'MDS output';
run;
proc print
    data=mysave.coord (keep=author dim1 dim2 dim3);
    var author dim1 dim2 dim3;
run;
options ps=**40**;

proc plot
    data=mysave.coord vtoh=**.5**;
  plot dim1*dim2 = '*' $ author / BOX
  HAXIS=BY **.4** VAXIS=by **1.3**;
  where _type_='CONFIG';
  plot (dim1 dim2) * dim3 = '*' $ author / BOX
  HAXIS=BY **.35** VAXIS=by **.35**;
  where _type_= 'CONFIG';
TITLE2 'PLOT Of CONFIGURATION';
run;

THE G3D PROCEDURE

The G3D procedure in the SAS system presents solutions in three dimensions using scatter plots, surface plots, and contour plots. The data can be represented as surfaces (surface plots), or as points (scatter plots). The three dimensional numeric variables can also be represented in two dimensions (contour plots). Of these types, ACA study results can be best displayed by scatter plots. The input to the G3D procedure is the coordinate matrix produced by the MDS procedure and saved as a permanent data set, mysave.coord.

SYNTAX OF THE G3D PROCEDURE

PLOT *plot-request* </option(s)>;

GENERATING A SIMPLE SCATTER PLOT

To generate a simple scatter diagram, the following proc g3d statement taken from SAS OnlineDoc Version 8 can be used as a starting point.

```
proc g3d data=reflib.iris;
    scatter petallen*petalwid=sepallen;
run;
```

The first line, proc g3d, specifies the data set name and starts the procedure. The second line contains the scatter statement for producing three-dimensional scatter plots using values of three numeric variables from the input data set. Using the input file (mysave.coord), the following PROC G3D statements create a simple 3D scatter plot.

```
libname mysave 'c:\wo\books\aca\sasdata\perm';
proc g3d data=mysave.coord;
```

```
scatter dim2*dim1=dim3;
run;
```

Figure 6.10 A Simple G3D Scatter Plot

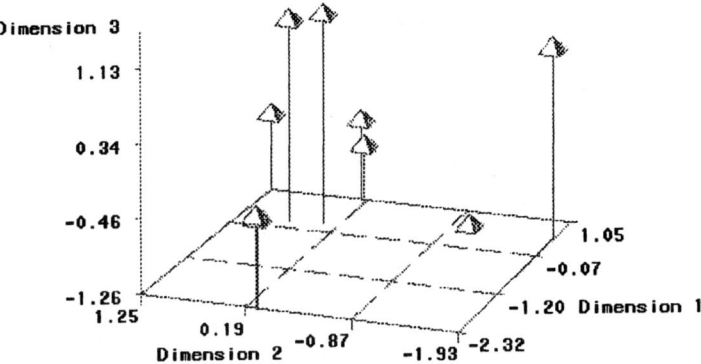

MODIFYING PLOTS WITH SCATTER OPTIONS

Plots can be modified with many SCATTER options. To show some of the options, the next PROC G3D statement adds options to add a grid and change the text of the axis labels.

```
libname mysave 'c:\wp\books\aca\sasdata\perm';
proc g3d data=mysave.coord;
scatter dim2*dim1=dim3
/grid caxis=black;
label dim1='foundations'
     dim2='model management'
     dim3='user-interface';
run;
```

130

Figure 6.11 A Scatter Plot with Options

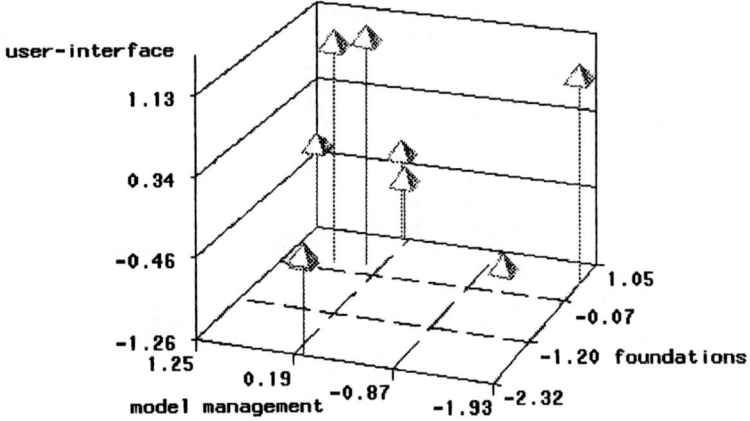

CREATING THE ANNOTATE DATA SET USING THE DATA STEP

In ACA study, 3D scatter plots without labels on data points provide little useful information for the ACA researchers. This section discusses how to label data points on a plot. Figures 6.12 and 6.13 show multiple data points with no labels. The annotate facility in the SAS system produce a figure to show the name of the author on each data point.

SAS PROGRAM TO CREATE THE ANNOTATE DATA SET (MYSAVE.ACAANNO)

The following program (Createanno.sas) creates the annotate data set, mysave.acaanno (Table 6.10).

Table 6.9 Creating an Annotate Data Set

```
libname mysave 'c:\wp\books\aca\sasdata\perm';
data mysave.acaanno1;
set mysave.coord;
length function $8;
retain   xsys ysys zsys '2'
         function 'label'
         size .4
         position '2'
         color 'blue'
         style 'zapf';
x=dim1;
y=dim2;
z=dim3;
text=author;
proc print;
run;
```

Table 6.10 An Annotate Data Set

```
The SAS System     12:03 Monday, January 20, 2003
Obs  _DIMENS_  _MATRIX_  _TYPE_      Author       _NAME_         Dim1

 1      3         .     CRITERION                                0.11468
 2      3         .     CONFIG      ALTER        ALTER           1.04522
 3      3         .     CONFIG      BONCZEK      BONCZEK        -2.31542
 4      3         .     CONFIG      CARLSON      CARLSON         0.17294
 5      3         .     CONFIG      HUBER        HUBER           0.47546
 6      3         .     CONFIG      KEEN         KEEN            1.01034
 7      3         .     CONFIG      SCOTTMORTON  SCOTTMORTON     1.00292
 8      3         .     CONFIG      SIMON        SIMON           0.71084
 9      3         .     CONFIG      SPRAGUE      SPRAGUE         0.17918
10      3         .     CONFIG      WHINSTON     WHINSTON       -2.28148
11      3         .     SLOPE                                    0.34472

Obs    Dim2      Dim3     function xsys ysys zsys size position

 1      .         .        label    2    2    2   0.4    2
 2    1.24929  -0.34354    label    2    2    2   0.4    2
 3    0.07483  -0.26530    label    2    2    2   0.4    2
 4    0.71272   1.04272    label    2    2    2   0.4    2
 5   -1.93306   0.88987    label    2    2    2   0.4    2
 6    0.25904  -0.30261    label    2    2    2   0.4    2
 7    0.22087  -0.58030    label    2    2    2   0.4    2
 8   -1.00424  -1.25772    label    2    2    2   0.4    2
 9    0.34683   1.13255    label    2    2    2   0.4    2
10    0.07373  -0.31568    label    2    2    2   0.4    2
11      .         .        label    2    2    2   0.4    2

Obs  color   style       x           y          z      text

 1   blue    zapf     0.11468       .          .
 2   blue    zapf     1.04522     1.24929   -0.34354   ALTER
 3   blue    zapf    -2.31542     0.07483   -0.26530   BONCZEK
 4   blue    zapf     0.17294     0.71272    1.04272   CARLSON
 5   blue    zapf     0.47546    -1.93306    0.88987   HUBER
 6   blue    zapf     1.01034     0.25904   -0.30261   KEEN
 7   blue    zapf     1.00292     0.22087   -0.58030   SCOTTMORTON
 8   blue    zapf     0.71084    -1.00424   -1.25772   SIMON
 9   blue    zapf     0.17918     0.34683    1.13255   SPRAGUE
10   blue    zapf    -2.28148     0.07373   -0.31568   WHINSTON
11   blue    zapf     0.34472       .          .
```

Understanding the SAS Program for Creating Annotate Data Set

Using the SAS program (Createanno.sas) and the annotate data (mysave.anno), this section discusses what each section of the program does by comparing it with the output (mysave.anno). The program produces the annotate data set which contains the observations to generate the output by using a DATA step.

The Data step begins with Data <Data Set Name>

DATA <DATA SET NAME>

/* Permaent Data Set Name and its location */
libname mysave 'c:\wp\books\aca\sasdata\perm';
DATA mysave.acaanno;

SET

The set statement reads an observation from one or more SAS data sets. Alternatively the data can be directly entered from the keyboard or the data may come from an infile format. Here, the set statement is one way of reading the data from the previous step, the MDS procedure in Figure 6.1.

SET<*SAS-data-set(s)* <(*data-set-options(s)*)>>
<*options*>;

The symbol <> should not be included in a SAS statement.

Example
set mysave.coord; or
set mysave.coord (keep=dim1 dim2 dim3 author);

To process the data set, mysave.coord, the keep= option may not be necessary. Table 6.5 shows that the data set only contains 5 columns of data with observation number. If there are numerous variables in the data set, only those variables that are listed after the KEEP= data set option are available for processing. Table 6.9 shows the annotated data set without using the KEEP= option. With the (keep=dim1 dim2 dim3 author) option, the annotated data set would exclude the four columns of data (_DIMENS_, _MATRIX_, _TYPE_, _NAME_).

LENGTH

The LENGTH statement creates a variable and set the length of the variable, as in the following example:

length function $8;

The length statement create a character variable, function, with maximum 8 characters. The length can be set any number of characters to hold the function name, LABEL. Therefore, it could be permissible to have $5 instead of $8. If the length is set to have a value of less than $5, the annotate facility will not produce annotations.

RETAIN

A retain statement lists all the variables that retain the same values for all observations in a annotated data set. Notice that in Table 6.9, the eight variables listed in the following retain statement hold the same values specified with single quotation.

```
RETAIN    xsys ysys zsys '2'
          function 'label'
          size .4
          position '2'
          color 'blue'
          style 'zapf';
```

COLOR **'color'** (the color of the text).

POSITION *'text-position'* | '0'; (the placement of the text string (author's name) in relation to the pyramid position of each author.

Text-position can be one of the following:
- Numbers 1 through 9,
- Characters A through F,
- Symbols <, +, or >.

Position '2' indicates the name of each author is to be printed at the location of one cell above the pyramid shape and it will be centered. (See Figure 6.14).

SIZE **height** specifies the height of the text string. The SIZE variable units are based on the value of the HSYS variable. In the above example, HSYS (the coordinate system) value is not specified. The default value is 4, which means the coordinate system unit is a cell in graphics output area. Size .4 means that the height of authors' name is 40% of the height of a cell in graphic area.

Style 'font style' the font style of the TEXT

FUNCTION

The purpose of the function statement is to specify a graphics command function for the Annotate facility to perform with the syntax, FUNCTION 'function-name';

The graphics command function includes BAR, CNTL2TXT, COMMENT, DEBUG, DRAW, DRAW2TXT, FRAME, LABEL, MOVE, etc. The default value is LABEL.

COORDINATE-SYSTEM

To produce a three dimensional graph, the value of the coordinate system (xsys, ysys, zsys) must be specified as '2'. The value of 'coordinate-system' are specified in Table 6.11.

ASSIGNMENT

Assignment statements in a DATA step evaluate the expression on the right side of the equal sign and store the result in the variable that is specified on the left side of the equal sign.

X=**horizontal-coordinate** is used to represent dimension 1
Y=**vertical-coordinate** is used to represent dimension 2
Z=**depth-coordinate** is used to represent dimension 3
Text=author (Text string values are from ID author names)

Table 6.11 The Value of The Coordinate System

Absolute Systems	Relative Systems	Coordinate System Units
1	7	percentage of data area
2	8	data values
3	9	percentage of graphics output area
4	A	cell in graphics output area
5	B	percentage of procedure output area
6	C	cell in procedure output area

Source: (SAS Institute Inc. 2000).

PROC G3D WITH THE *ANNOTATE* =*OPTION*

Libname mysave 'c:/wp/books/aca/sasdata/perm'
PROC G3D data=mysave.coord;
SCATTER dim2*dim1=dim3
/anno=mysave.acaanno grid caxis=black
shape='pyramid';
RUN;

Table 6.12 PROC G3D Procedure with Temporary Data Sets

```
data ACA9;
INPUT Author $ 1-12 @13 x1-x9;
CARDS;
ALTER         121   25   66   23   92   83   38   67   27
BONCZEK        25  103   53   23   46   42   33   59   93
CARLSON        66   53  173   34  112  101   54  133   54
HUBER          23   23   34   68   47   39   37   49   24
KEEN           92   46  112   47  206  174   82  126   50
SCOTTMORTON    83   42  101   39  174  190   82  105   45
SIMON          38   33   54   37   82   82  111   58   37
SPRAGUE        67   59  133   49  126  105   58  182   61
WHINSTON       27   93   54   24   50   45   37   61  104
;
%include 'c:\program files\SAS Institute\SAS\V8\stat\sample\xmacro.sas';
%include 'c:\program files\SAS Institute\SAS\V8\stat\sample\distnew.sas';
%distance ( data=aca9,
        id=author,
        options=print,
        method=dcorr,
        var=x1-x9,
        out=dist);
proc print data=dist;
run;

/**** MDS procedures to produce profiles****/
proc mds data=dist
    condition=un
    level=ratio
    coef=identity
    dimension=3
    cutoff=0
    fit=2
    pcoef
    pconfig
    pfinal
    out=coord
    outres=res;
    ID author;
;
title1 'MDS output';
run;
proc print data=coord (keep=author dim1 dim2 dim3);
var author dim1 dim2 dim3;
run;
```

```
options ps=40;

proc plot data=coord vtoh=.5;
   plot dim1*dim2 = '*' $ author / BOX
     HAXIS=BY .4 VAXIS=by 1.3;
    where _type_='CONFIG';
   plot (dim1 dim2) * dim3 = '*' $ author / BOX
     HAXIS=BY .35 VAXIS=by .35;
    where _type_ = 'CONFIG';

     TITLE2 'PLOT Of CONFIGURATION';
data anno;
set coord (keep=dim1 dim2 dim3 author);
length function $8;
retain xsys ysys zsys '2'
      function 'label'
      size .4
      position '2'
      color 'blue'
      style 'zapf';
x=dim1;
y=dim2;
z=dim3;
text=author;
proc g3d data=coord;
scatter dim2*dim1=dim3 /anno=anno shape='pyramid';
run;
quit;
```

Table 6.13 PROC G3D Statement with Permanent Data Sets

```
DATA ACA9var;
INFILE 'C:\wp\books\aca\sasdata\9varname.dat';
INPUT Author $ 1-15 @17 x1-x9;
;
libname mysave 'c:\wp\books\aca\sasdata\perm';
%include 'c:\program files\SAS Institute\SAS\V8\stat\sample\xmacro.sas';
%include 'c:\program files\SAS Institute\SAS\V8\stat\sample\distnew.sas';
%distance (data=aca9var,
           id=author,
           options=print,
           method=dcorr,
           var=x1-x9,
           out=mysave.dist);
proc print data=mysave.dist;
run;

/**** MDS procedures to produce profiles****/
proc mds   data=mysave.dist
           condition=un
           level=ratio
           coef=identity
           dimension=3
           cutoff=0
           fit=2
           pcoef
           pconfig
           pfinal
           out=mysave.coord
           outres=res;
           ID author;
;
title1 'MDS output';
run;
proc print data=mysave.coord (keep=author dim1 dim2 dim3);
var author dim1 dim2 dim3;
run;
options ps=40;

proc plot  data=mysave.coord vtoh=.5;
           plot dim1*dim2 = '*' $ author / BOX
           HAXIS=BY .4 VAXIS=by 1.3;
           where _type_='CONFIG';
           plot (dim1 dim2) * dim3 = '*' $ author / BOX
```

```
                HAXIS=BY .35 VAXIS=by .35;
                where _type_ = 'CONFIG';

        TITLE2 'PLOT Of CONFIGURATION';
data mysave.anno;
set mysave.coord (keep=dim1 dim2 dim3 author);
length function $8;
retain      xsys ysys zsys '2'
            function 'label'
            size .4
            position '2'
            color 'blue'
            style 'zapf';
x=dim1;
y=dim2;
z=dim3;
text=author;
PROC PRINT;
PROC G3D data=mysave.coord;
scatter dim2*dim1=dim3 /grid caxis=black anno=mysave.anno shape='pyramid';
run;
quit;
```

Figure 6.12 Scatter (Dim1*Dim2) = Dim3

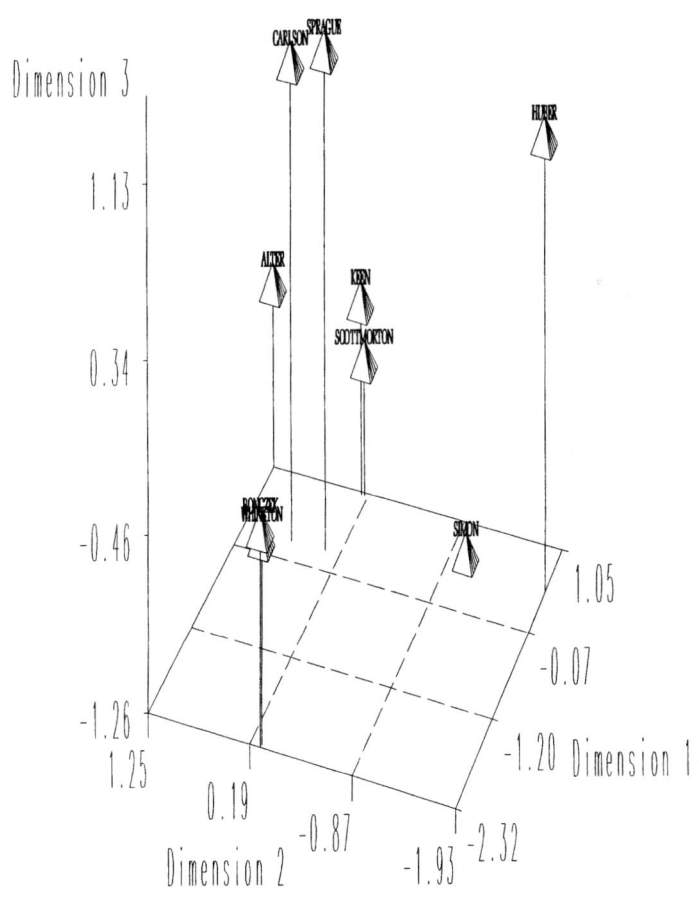

CHAPTER 7 AN ACA STUDY IN AN INFORMATION SYSTEMS AREA: THE INTELLECTUAL STRUCTURE OF DECISION SUPPORT SYSTEMS RESEARCH (1969-1990)

This chapter[1] is to infer the intellectual structure of the DSS field by means of an empirical assessment of the DSS literature over the period 1969 through 1990. Three multivariate data analysis tools (factor analysis, multidimensional scaling, and cluster analysis) are applied to an author cocitation frequency matrix derived from a large database file of comprehensive DSS literature over the same period. Seven informal clusters of decision support systems (DSS) research subspecialties and reference disciplines were uncovered. Four of them represent DSS research subspecialties-- foundations, group DSS, model/data management, and individual differences. Three other conceptual groupings define the reference disciplines of DSS-- organizational science, multiple criteria decision making, and artificial intelligence. DSS is a very young academic field and is still growing. DSS has just entered the era of growth after 20 years of research. During the 1990s, DSS

---

1 This chapter is largely based on the two earlier publications from OMEGA, Vol.23, No.5, 1995, Pages 511-523, Eom, Decision support systems research: Reference disciplines and cumulative tradition, Copyright 1995 and Journal of the American Society for Information Science, Vol. 47, No.12, December 1996, Eom and Farris, "The Contributions of Organization Science to the Development of Decision Support Systems Research Subspecialties with Permission from Elsevier Science and Wiley

research will be further grounded in a diverse set of reference disciplines. Furthermore, DSS is in the active process of solidifying its domain and demarcating its reference disciplines. A DSS theory is imminent in the very near future in some areas of DSS research such as model management.

INTRODUCTION

Within the managerial and organizational context, it was in the early 1950s when one of the first computers began to process payroll data. Since then, the study of computers and information systems has evolved continuously. Since the early 1970s, scholars in the management information systems (MIS)/decision support systems (DSS) areas have recognized the important roles computer-based information systems play in supporting managers in their semi-structured or unstructured decision making activities. For example, Gorry and Scott Morton (1971, p. 57) made the controversial claim that "Information systems should exist only to support decisions." Since then, there has been a growing amount of research in the area of DSS (Elam et al. 1986; Eom et al. 1990a; Farhoomand 1987; Teng et al. 1990). As Keen (1980) indicated in the early 1980s, it is necessary for information systems research to clarify reference disciplines and to build a cumulative tradition to become a coherent and substantive field. This is necessary for DSS research as well. In the DSS area, Eom, Lee, and Kim (1993b) conducted an initial study to identify two areas of contributing disciplines (management science and multiple criteria decision making) and five subspecialties of DSS research (foundations, group DSS, database management systems, multiple-criteria DSS, marketing DSS, and routing DSS). Due to the restrictive nature of their data set (specific DSS applications only), their study failed to provide a comprehensive picture of DSS research subspecialties.

Several recent surveys have reported a macro-view of decision support systems development status and their evolution as an academic field. A study by Farhoomand (1987) reports that DSS has been one of the five top research themes and has shown steadily increasing acceptance among information systems researchers in the last nine years (1977-1985). A recent survey, based on perceptions of a sample of MIS researchers, reports that almost one third of the respondents were conducting DSS research (Teng et al. 1990). The results of another survey indicate that an increasing number of decision support systems have been developed and implemented in many for-profit and not-for-profit organizations in every functional area of decision making (Eom et al. 1990b). All these research efforts indicate that the fields of DSS have become an important part of computer-based information systems.

Decision support systems are a relatively young field of study, compared, for example, to economics, organizational behavior, etc. As a field of study continues to grow and becomes a coherent field, study of the intellectual development of the field is important. Culnan (1986, p.156) stressed its importance in this way: "Researchers can benefit by understanding this process and its outcomes because it reveals the vitality and the evolution of thought in a discipline and because it gives a sense of its future. In a relatively new field such as MIS, this understanding is even more beneficial because it identifies the basic commitments that will serve as the foundations of the field as it matures." For DSS to become a coherent and substantive field, a continuing line of research must be built on the foundation of previous work. Without it, there may be good individual fragments rather than a cumulative tradition (Keen 1980). When speaking about a cumulative research tradition, Keen (1980, p.18) emphasized the importance of reflecting upon research that has already been done: "It atrophies if it cuts itself off from curiosity, diversity, and reflection" and "Let us make sure we keep a few philosophers, historians, general systems theorists and social activists *within* our network; even if only to write useful survey papers."

A number of prior studies have been conducted to assess the extent of progress within these stages in the information systems area. Culnan (1986; 1987) and Cushing (1990) conducted examinations of the intellectual evolution and development of the MIS area. These authors concluded that significant progress had been made toward a cumulative research tradition in MIS and identified several groups of MIS research subfields. On the other hand, others perceived that MIS researchers felt that there was an overemphasis on transient topics and that there was continuing evidence of fragmentation and lack of both a cumulative research tradition and articulated MIS theories (Farhoomand 1987; Teng et al. 1990). Thus, there have been conflicting assessments as to the existence of a cumulative research tradition in the MIS area and the evolution of MIS as a field of scientific research. In the DSS area, Eom, Lee, & Kim (1993b) conducted the first study which identified the intellectual structure, cumulative tradition, and reference disciplines of DSS. Their study identifies several subspecialties of the DSS area, such as foundations, group DSS, multiple criteria DSS, marketing and routing DSS, and database management systems. However, several active DSS research areas, such as model management systems and interface systems were not represented by their study. Consequently, they concluded that the DSS field has failed to build a cumulative research tradition and that a DSS cumulative research tradition does not exist. Furthermore, they claim that DSS research over the past two decades has failed to develop its own DSS theory. The important factor contributing to their conclusions seems to be the limited nature of their bibliographic databases which included only specific DSS application articles (Eom et al. 1990b) and excluded a large number of articles which deal with the theoretical development of decision support systems.

This study attempts to overcome the weakness of the study of Eom *et al.* (1993b) by expanding the number of citing articles from 259 application articles to 632 articles, thereby fostering a better understanding of how DSS has evolved to its present state. Using three multivariate analyses (factor, cluster, and multidimensional scaling) of author cocitation matrix, this study attempts to identify the intellectual structure, reference disciplines, and major themes in current DSS research and provide important groundwork for future theoretical development and scientific inquiry in the DSS area.

DATA

The primary data for this study were gathered from a total of 692 citing articles in the DSS area over the past 20 years (1969-1990). These citing articles consist of two categories--270 specific DSS application articles based on the same definition and collection criteria of Eom and Lee (1990b) and 422 non-application articles. Of these 270 specific DSS application articles, there are 259 overlapping articles between this study and the previous study (Eom et al. 1993b). In other words, this study is based on a database file consisting of 259 citing articles from the previous study (Eom et al. 1993b) as well as 11 additional specific DSS articles and 422 non-application articles. These 692 articles were collected from the following three sources: 210 articles over the period of 1975 through 1985 from (Elam et al. 1986); 157 articles over the period of 1969 through 1987 from (Sprague et al. 1989); 203 articles over the period of 1969 through 1988 from (Eom et al. 1990b); and 220 additional articles (included to cover the period the other articles did not cover) taken from the same source journals and selected using the same selection criteria used by the three source articles. See Chapter 3 for further details on the selection criteria for citing articles. There was an overlap of 98 citing articles in the three sources, and only those unique articles were added to our database. Subsequent steps were taken to create a database file that consists of a total of 15,030 *cited* reference records taken from the *citing* articles.

RESEARCH METHODOLOGY

This study is based on the assumptions that "bibliographic citations are an acceptable surrogate for the actual influence of various information sources" (McCain 1984) and that the cocitation analysis of a field yields a valid representation of the intellectual structure of the field (Bellardo 1980; McCain

1984; McCain 1986; Smith 1981).

Three multivariate data analysis tools (factor analysis, multidimensional scaling, and cluster analysis) were applied to an author cocitation frequency matrix derived from a large database file of comprehensive DSS literature over the period of 1969 through 1990. The primary research method for this study, Author Cocitation Analysis, consists of the following steps.

SELECTION OF AUTHOR

Authors for this study were selected by a two-stage process. The first step is the loose screening based on the minimum citation of 60 or more times during the period of 1969-1990, based on the prior work of Culnan (1986). This stage yields a list of about 100 names.

COMPILATION OF COCITATION FREQUENCY MATRIX

To overcome a standard problem with the Institute for Scientific Information (ISI) databases which code only the first author of a cited work, a Fox-Base based matrix generation system was developed to compute a cocitation frequency between any pair of authors. The cocitation matrix generation system we developed gives access to cited coauthors as well as first authors.

In the next stage, the final author set was determined. Due to the possible instability of small cocitation counts, author cocitation analysis researchers introduced several ad hoc criteria for further screening a large pool of candidate authors to finalize a list of authors. The criteria include a *mean cocitation rate* above a certain lower limit per author in each time period (e.g., nine for 10 years of Social Scisearch data), cocitation with at least one-third of the entire author set, or restricting the final author set to the 20% receiving the highest number of citations and cocitations in initial retrieval trials. For further details on several different approaches to compiling a predetermined list of authors, see McCain (1990). However, all these criteria were suggested to be applied to the commercial on-line databases such as SCISEARCH and SOCIAL SCISEARCH.

Our databases are significantly different from those commercial databases in terms of size of records. Besides, the cocitation matrix generation system we developed gives access to cited coauthors as well as first authors. Due to these differences, we could not follow the suggested criteria e.g., (McCain 1990) such as "a mean cocitation rate of 'x' or more cocitations in each time period." Rather, we had to invent a new criterion through the method of trial and error. We experimented with the sensitivity of changing the cocitation threshold on the final outcomes (number of meaningful factors to accurately represent DSS research subspecialties). With our databases, we conclude that the number of cocitations of an author with himself/herself can be a better criterion to determine the final author set due to the simplicity. Applying the mean cocitation criteria to our Lotus worksheet file (the output from the cocitation matrix generation system) involves too many computations. Whenever we delete/add an author to the final author set, we need to compute the mean cocitation rate of each author again. Using the cocitation rate of 25 with himself/herself in the period (1969-1990), the initial set of 100 authors was reduced to 67 as the final author set for further analysis. Later, we experimented to lower the threshold from the 25 cocitation rate to a 20 cocitation rate. Although the number of authors increased from 67 to 80, the number of meaningful factors remained the same, and the major conclusions reported here are not changed. Also, it is important to point out that cocitation thresholds themselves, as sole connection criteria, are suspect in a highly multidisciplinary area. One should look at the overall connectedness and the focused cocitation counts as well.

DATA ANALYSIS

The raw cocitation matrix of 67 authors was analyzed by principal components analysis with the latent root criterion (eigenvalue 1 criterion) applied to obtain the initial solution of 9 factors (factor loadings greater than .40). In principal components analysis, only those factors with eigenvalues (the column sum of squares for a factor) greater than 1 are interpreted as significant. Out of the two major rotation options, we chose an oblique rotation method. The PROMAX rotation specification provides both orthogonal and oblique rotations with only one invocation of PROC FACTOR. Compared to an orthogonal rotation method, the oblique factor rotation is "more desirable because it is theoretically and empirically more realistic" (Hair et al. 1987, p.245). It allows a more natural rotation without the imposition of orthogonal factors. Moreover, it generates additional information about the correlations between the factors (Table 7.2). The nine extracted factors account for 84.66 percent of the total variance of the data set.

The correlation matrix derived from the author cocitation matrix was used as an input to the multidimensional scaling program PROC MDS of personal computer SAS (release 6.8) to visualize the similarity and dissimilarity within each group of DSS researchers, as well as the similarity and dissimilarity among the various groups of DSS researchers. The correlation matrix was generated by the %DISTANCE macro (updated on June 28, 1994) of the SAS/STAT sample library of SAS Institute Inc. The %DISTANCE macro computes various measures of proximity and stores them as a lower triangular matrix in an output data set. The output data set is used as input to the MDS and CLUSTER procedures (a hierarchical agglomerative clustering program with Ward's trace option). PROC MDS procedure includes many of the features of the ALSCAL procedure (Young et al. 1986) and some features of the MLSCALE procedure (Ramsay 1986) (SAS Institute Inc. 1992).

RESULTS OF FACTOR ANALYSIS

Table 7.1 presents the nine factors extracted from the correlation matrix using principal components analysis. Care must be exercised when interpreting the substantiality of factors 7, 8, and 9 which include only two authors, no loadings above 0.6, and in which the authors also load on other factors with the same or higher absolute values. The uncovered DSS research subspecialties are foundations (factor 1), group DSS (factor 2), model management (factor 3), user interface/individual differences (factor 4), and implementation (factor 8). Four other conceptual groupings define the reference disciplines of DSS--organizational science (factor 5), multiple criteria decision making (MCDM) (factor 6), strategic management (factor 7), and group decision making (factor 9).

Table 7.1 Rotated Factor Correlations Matrix (1969-1990)

Factor 1 Foundations		Factor 2 GDSS		Factor 3 Model Management	
Keen	0.96	Gallupe	0.95	Whinston	0.94
Scott Morton	0.96	Hiltz	0.92	Bonczek	0.94
Alter	0.92	Gray	0.91	Holsapple	0.93
Carlson	0.89	Bui	0.91	Lanning	0.91
Sprague	0.89	Turoff	0.91	Elam	0.90
Gorry	0.84	Kraemer	0.91	Henderson	0.86
Ginzberg	0.83	Applegate	0.90	Stohr	0.86
Little	0.82	King, J.L.	0.90	Dolk	0.83
Simon	0.82	DeSanctis	0.87	Geoffrion	0.79
Bennett	0.81	Nunamaker	0.86	Davis, R.	0.75
Anthony	0.81	George	0.81	Konsynski	0.73
King, W.	0.79	Huber	0.77	Sprague	0.68
Watson, H.	0.79	Jarke	0.71	Shortliffe	0.67
Wagner, G.R.	0.78	Delbecq	0.68	Carlson	0.65
Meador	0.77	Van Den Ven	0.68	Naylor	0.65
Davis, G.B.	0.75	Konsynski	0.67	Watson, H.	0.62
Rockart	0.73	Jarvenpaa	0.58	Bennett	0.57
Naylor	0.72	Jelassi	0.51	Courtney	0.57
Courtney	0.65	Dickson	0.46	Meador	0.55
Robey	0.62			Simon	0.53
Ackoff	0.59			Keen	0.52
Zmud	0.58			Jarke	0.52
Stohr	0.57			Scott Morton	0.50
Lucas	0.56			Anthony	0.47
Mintzberg	0.56			Jelassi	0.46
Henderson	0.55			Ginzberg	0.42
Sanders, G.	0.55			Huber	0.41
Chervany	0.54				
Shortliffe	0.53				
Davis, R.	0.50				
Whinston	0.50				
Holsapple	0.49				
Mason, R.	0.48				
Benbasat	0.48				
Newell	0.48				
Bonczek	0.48				
Mitroff	0.44				
Huber	0.43				
Blanning	0.42				
Variance	19.28		12.63		14.94
% Variance	28.77		18.85		22.29

Continued on next page

Table 7.1---Continued

Factor 4 User Interface		Factor 5 Org. Science		Factor 6 MCDM	
Benbasat	0.95	March	0.84	Keeney	0.94
Dexter, A.	0.92	Simon	0.84	Raiffa	0.93
Lusk	0.90	Newell	0.82	Geoffrion	0.66
Lucas	0.90	Mintzberg	0.75	Jelassi	0.60
Dickson	0.88	Tversky	0.74	Jarke	0.53
Zmud	0.87	Ackoff	0.68	Stohr	0.43
Ives	0.86	Anthony	0.56		
Chervany	0.85	Scott Morton	0.54		
Robey	0.73	Keen	0.54		
Jarvenpaa	0.67	Davis, G.B.	0.55		
Mason, R.	0.61	Gorry	0.50		
Mitroff	0.61	Mitroff	0.49		
King, W.	0.57	Courtney	0.49		
Simon	0.56	Davis, R.	0.48		
Davis, G.B.	0.55	Shortliffe	0.48		
Keen	0.54	Mason, R.	0.47		
Courtney	0.53	Huber	0.46		
Scott Morton	0.52	Sprague	0.46		
DeSanctis	0.52	Meador	0.45		
Mintzberg	0.51	Sanders	0.44		
Sanders	0.48	Carlson	0.44		
Ginzberg	0.47	Benbasat	0.43		
Huber	0.45	Zmud	0.43		
Newell	0.43	Lucas	0.42		
Gorry	0.43	Robey	0.41		
Ackoff	0.41	Dickson	0.40		
Sprague	0.40				
Variance	10.96		4.42		2.77
% Variance	16.35		6.59		4.14

Factor 7 Strategic Mngt		Factor 8 Implementation		Factor 9 Group DM	
Mitroff	0.65	Sanders	0.68	Van Den Ven	0.46
Mason	0.63	Courtney	0.64	Delbecq	0.45
Jelassi	0.44	Henderson	0.51	Nunamaker	0.41
Ackoff	0.43	Huber	0.45	Davis, G.	0.40
		Zmud	0.43		
		Konsynski	0.42		
		Robey	0.41		
		Elam	0.40		
Variance	2.77		5.07		4.42
% Variance	4.14		7.57		4.87

Figure 7.1 Major Factor Correlation Network

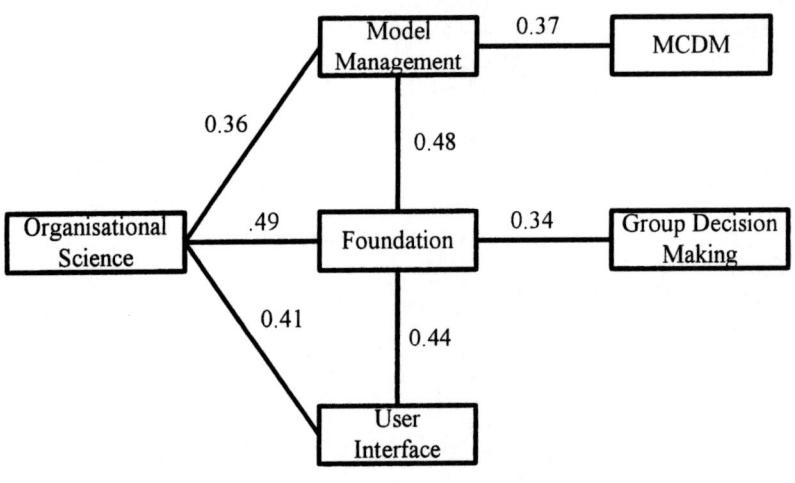

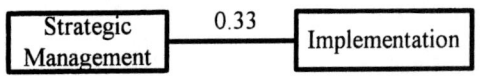

REFERENCE (CONTRIBUTING) DISCIPLINES

There seem to be several contributing disciplines that have influenced the evolution of the current state of DSS research subspecialties. There have been a number of assumed reference disciplines in the DSS area such as psychology, economics, computer science, political science, etc. Nevertheless, this study identified only weak influence from organizational sciences, artificial intelligence, and multiple criteria decision making on the development of DSS research subspecialties.

<u>ORGANIZATIONAL SCIENCES</u>

Factor 5 appears to represent *Organizational Sciences*. DSS are designed and implemented to support organizational as well as individual decision making. Without a detailed understanding of decision making behavior in organizations, "decision support is close to meaningless as a concept" (Keen et al. 1978, p. 61). Organizational scientists have classified organizational decision making in terms of several schools of thought: 1) the rational model focusing on the selection of the most efficient alternatives, with the assumption of a rational, completely informed, single decision maker; 2) the organizational process model by Cyert and March (1963) stressing the compartmentalization of the various units in any organization; 3) the satisficing model by Simon and his colleagues (Newell et al. 1972; Simon 1969) to find an acceptable, good enough solution, reflecting "bounded rationality"; 4) and other models.

<u>MULTIPLE CRITERIA DECISION MAKING</u>

Factor 6 seems to represent *multiple criteria decision making (MCDM)*. MCDM deals with semistructured and unstructured decisions involving multiple attributes, multiple objectives, or both. As reported by Dyer and others (1992), numerous individuals have contributed to give rise to the field of MCDM. Among them, Keeney and Raiffa (1976) have provided us with an excellent and complete overview of multiple attribute utility theory, along with numerous examples of practical applications. By the nature of multiple criteria decision making, usually there are numerous nondominated solutions in MCDM problems. To single out a decision alternative, Geoffrion, Dyer, and Feinberg (1972) suggested interactive procedures for multiple criteria optimization. Since the mid-1980s, MCDM model-embedded decision support systems, represented by Jelassi, Jarke and Stohr (1985) are emerging to forge a new branch of DSS research. Factor 7, defined by Mason and Mitroff, seems to represent *Strategic Management*. Despite numerous planning frameworks and methodologies, our cocitation analysis shows that the assumption surfacing methodology suggested by Mason and Mitroff has been an influential framework for addressing the full range of management support systems' requirements.

GROUP DECISION MAKING

Factor 9, *Group Decision Making*, is represented by Delbecq and Van De Ven. Their research in the 1970s experimentally compared three alternative methods for group decision making: the conventional interacting (discussion) group, the nominal group technique, and the Delphi technique.

DSS RESEARCH SUBSPECIALTIES

In 1980, Keen (1980) stated that MIS research lacked a cumulative tradition. In his view, there was virtually no cumulative research tradition in the MIS area without "continued follow-up on interesting lines of inquiry." Several areas of DSS research subspecialties that emerged in this study provide us with some evidence as to the existence of a cumulative research tradition.

FOUNDATIONS

Most authors in factor 1 conducted descriptive research to provide definitions and concepts in the very early stage of DSS development. Some authors clearly pinpointed a need for another type of information systems to relieve managers' suffering from an "over abundance of irrelevant information" (Ackoff 1967) and therefore Gorry and Scott Morton (1971) claimed that "Information systems should exist only to support decisions." Anthony (1965) classified all managerial activities into three categories: strategic planning, management control, and operational control. This taxonomy combined with that of Simon (1960) which classified all decisions into structured, semistructured, and unstructured provided a simple schema for classifying organizational decisions to be best supported by TPS, MIS, and DSS. Little (1970) suggested a concept of *decision calculus* as "a model-based set of procedures for processing data and judgments to assist a manager in his decision making." Although he did not use the term DSS, he proposed the concept of a decision calculus which has several desirable characteristics of DSS (*simple, robust, easy to control, adaptive, complete on important issues, easy to communicate with*). Sprague and Watson (1975) examined the necessity of including decision models in an integrated MIS and emphasized that there is a need for a systematic way of embedding decision support models into MIS to support managers' decision making processes.

Keen and Scott Morton (1978) extended these previous works and suggested a widely accepted definition of DSS which implies "the use of computers to: assist managers in their decision processes in semistructured tasks; support, rather than replace managerial judgment; and improve the effectiveness of decision making rather than its efficiency. Keen and Scott Morton (1978) suggested three important areas of DSS research from an organizational perspective: design, implementation, and evaluation of DSS. Sprague and Carlson (1982) added several important further research areas --- data, model, dialogue, and decision makers, which can be termed DSS architecture. In addition, Sprague (1980) suggested an important and widely accepted definition and concept termed specific decision support systems.

Through the analysis of 56 implemented specific DSS, Alter (1975; 1977; 1980) classified all DSS into seven distinct types and added several folders into the implementation drawers: patterns, risk factors, and strategies of DSS implementations. In the early 1980s, Wagner (1981, p. 77) maintained that "If the DSS concept has a valid core, it must be secured against adulteration and overburdening by *evidence drawn from actual practice*" and the valid core of the DSS concept is to provide "interactive support for the thought processes of one or more executives in their principal function of making decisions." Others suggested theoretical models. William King and Rodriguez (1978) suggested an evaluation process model for evaluating MIS and DSS, which measures attitude, value perception, information usage, and decision performance in every stage of the system development life cycle in a simulated environment.

Several of these authors began to conduct empirical studies. Among them, Ginzberg's earlier work (1981), based on an empirical test of the level-of-adoption hypothesis, suggested that if full benefit is to be realized, DSS must be used as a catalyst for changes in the definition of the manager's role and DSS should be viewed in the broader context of organization change, and therefore the design of DSS is likely to be more successful if it incorporates 1) user participation, 2) normative system modeling, and 3) evolutionary or iterative design. Sanders and Courtney (1985) reported the results of a field study of organizational factors to ascertain the influence of success factors (the decision context, task interdependence, and task constraints) of DSS implementations.

GDSS

Since the mid-1980's we have witnessed an emerging DSS research theme: group decision support systems (factor 2). Earlier works by Delbecq, Van de Ven, and Gustafson (1975) experimentally compared three alternative methods for group decision making: the conventional interacting (discussion) group, the nominal group technique, and the Delphi technique. Many of these techniques (silent and independent idea generation, presenting each idea in a round-robin procedure, silent independent voting, etc.) were successfully utilized in the development of GDSS in the 1980's. Turoff and Hiltz (1982) conducted two experiments to study the impact of computer-based conferencing systems on group decision making and concluded that GDSS helped the computer-aided groups reach quality decisions more often than groups unaided by a GDSS. In the early stages of GDSS development, several descriptive research papers have been cornerstones for subsequent GDSS empirical research. Huber (1984) provided a comprehensive definition and proposed an architecture of GDSS. Further, alternative GDSS design strategies were examined to conclude that an activity-driven design strategy is superior to either a task-driven or techniques-driven strategy. His analysis of group activities led to another important conclusion that textual and relational information (PERT network, or organizational chart) is relatively more important for GDSS than it is for single user DSS. Another landmark paper is the result of the work of DeSanctis and Gallupe (1987); it presents an overview of GDSS, the potential impact of GDSS on group processes and outcomes, and proposes a multidimensional taxonomy of GDSS, based on the four dimensions: group size (smaller, larger); member proximity (dispersed, face-to-face); task type (6 types); and GDSS tool type (levels 1, 2, and 3). Kraemer and King, J.L. (1988) presents a comprehensive assessment of GDSS development and use in the U.S. by reviewing the current status of GDSS activities. They conceive GDSS as a sociotechnical "package" of (1) hardware, (2) software, (3) organizationware and (4) people; They classified GDSS into the following 6 types: The Electronic Boardroom, The Teleconferencing Facilities, The Information Center, The Decision Conference, The Collaboration Laboratory, and The Group Network.

During the second half of the 1980's, a group of researchers began to conduct empirical GDSS research. There are four comprehensive reviews of major GDSS research (Benbasat et al. 1993; Dennis et al. 1993; Dennis et al. 1988; Pinsonneault et al. 1989). Dennis *et al.* (1988) identified at least four streams of research under the broader label of experimental GDSS research to compare: Local Area Decision Nets (LADNs) to Decision Rooms, LADNs to no computer support, Decision Rooms to no computer support, and two different configurations of the same Decision Room. Gallupe, DeSanctis, and Dickson

(1988) added one more value of task type (group problem finding) to the dimension III of the GDSS cabinet and conducted an empirical investigation of group problem finding (smaller group, face-to-face, level 1, problem finding). Jarke, Jelassi, and Bui seem to define an important field of GDSS -- multiple criteria decision making (MCDM)-model embedded group decision support systems (Bui et al. 1984). The next subgroup includes Nunamaker (1987), Applegate, George (Dennis et al. 1988), and Konsynski (1984-1985) of the electronic meeting systems research. The taxonomy of EMS environments presented by Dennis *et al.* (1988) added a new time dimension (dimension VI: synchronous and asynchronous meetings) and another value (multiple group sites) to the dimension II of the DeSanctis and Gallupe taxonomy.

MODEL/DATA MANAGEMENT

Data/Model Management Systems have emerged as the third major research area (factor 3) in the DSS field. Since 1975, model/data management has been researched to encompass several central topics such as model base structure and representation, model base processing, and application of artificial intelligence to model integration, construction, and interpretation (Chang et al. 1993). In the model structure and representation area, the structured modeling approach by Geoffrion (1987) has significantly advanced the model representation area of model management, which is a significant extension of the entity-relationship data model and a necessary step for advancing to the next stage of model management (model manipulation). Dolk and Konsynski (1984) developed the model abstraction structure for representing models as a feasible basis for developing model management systems. Dolk attempts to connect both artificial intelligence and database management to evolve a theory of model management via model integration relied heavily upon the relational database theory.

In the model processing area, Blanning (1982) investigated important issues in the design of relational model bases and presented a framework for the development of a relational algebra for the specification of join implementation in model bases. In the area of AI application to model management, Bonczek, Holsapple, and Whinston (1979; 1980a; 1980b; 1981) suggested the use of artificial techniques for determining how models and data should be integrated in response to a user query. Elam, Henderson, and Miller (1980) introduced the concept of knowledge-based model management systems (MMS) to support a variety of complex decision problems with the use of semantic nets. They contended that the knowledge-based MMS could facilitate the use of the analytical tools in structuring as well as analyzing decision problems. Although model management research has not progressed enough to develop a theory of models, Dolk and Kottemann (Dolk et al. 1993, p. 51) believe that the emergence of a theory of models is imminent and the current model integration research is projected as "the springboard for building a theory of models equivalent in power to relational theory in the database community."

Dolk and Kottemann (1993) further believe that model management needs to see some effective implementations, much like relational theory needed ORACLE and other commercially viable products. The expense of building such systems is high, however, and it is not clear that there is market support for such a product. It is hoped that someone will achieve a breakthrough in this regard. Comprehensive literature reviews on model management can be found in (Blanning 1993; Chang et al. 1993).

USER INTERFACE/INDIVIDUAL DIFFERENCES

The initial investigation of user interface/individual differences (factor 4) was begun by the earlier works of Mason and Mitroff (Mason et al. 1973, p. 478), who hypothesized that "What is information for one type will definitely not be information for another. Thus, as designers of MIS, our job is not to get (or force) all types to conform to one, but to give each type the kind of information he is psychologically attuned to and will use most effectively." Bariff and Lusk (1977) presented a model for useful classification of behavioral variables for attaining successful MIS design. The Bariff and Lusk model proposed that the successful design and implementation of an MIS should explicitly involve consideration of the system's user cognitive styles. Benbasat and Dexter (1979; 1982) conducted a series of similar experiments to conclude that "an appropriate information system design can help overcome a mismatch between task environment and psychological type" (Benbasat et al. 1982, p. 8). Despite those positive claims emphasizing the user's cognitive style as an important consideration in the design of management information systems and DSS, Huber (1983, p. 567) concluded that "(1) the currently available literature on cognitive styles is an unsatisfactory basis for deriving operational guidelines for MIS and DSS designs." (2) "Further cognitive style research is unlikely to lead to operational guidelines for MIS and DSS designs."

Other subgroups of researchers in this factor have focused on the evaluation of graphical and color enhanced information presentation and other presentation formats (e.g., Tabular). They include Chervany and Dickson (1974) and Dickson, *et al.* (1977): comparison of the decision impacts of detailed reports with summarized reports; Lusk and Kersnick (1979), Lucas and Nielson (1980), and Lucas (1981): comparison of tabular with graphics; and DeSanctis (1984) comprehensively investigated previous research in this area up to 1984. Despite the numerous previous research reports, results are confusing and inconclusive: "the extravagant claims favoring graphic presentation formats may be considerably overstated" (Ives 1982, p. 21), "a picture may *not* be worth a thousand words" (DeSanctis 1984, p. 482), and there are no differences between tabular and graphical reports in terms of decision quality (Benbasat 1974). Jarvenpaa, Dickson, and DeSanctis (1985) argued that numerous equivocal findings could be attributable to the various tasks used in these experiments and the match between the task and presentation method as well as the lack of a sound taxonomy for classifying data extraction tasks and recommended the development of some type of taxonomy of tasks as a basis of interpreting the impact of the graphical presentation format. Tan and Benbasat (1993) have provided taxonomies for classifying various tasks and for classifying information presentation methods and concluded that "the task and presentation notion of matching provides a way to explain the conflicting results by showing that information presentation methods can not be evaluated outside the given task context in which they are applied" (Tan et al. 1993, p. 168).

DSS IMPLEMENTATION

Sanders and Courtney (1985) reported the results of a field study to ascertain the influence of success factors (the decision context, task interdependence, and task constraints) on DSS implementations (factor 8).

Table 7.2 Interfactor Correlations (1970-1990)

	Factor 1	Factor 2	Factor 3	Factor 4	Factor 5	Factor 6	Factor 7	Factor 8	Factor 9
Factor1	1.00000								
Factor2	-0.08806	1.00000							
Factor3	0.47779	0.13428	1.00000						
Factor4	0.43634	0.08420	0.11169	1.00000					
Factor5	0.48577	0.13666	0.36389	0.41263	1.00000				
Factor6	0.10256	0.15125	0.31792	0.08737	0.14409	1.00000			
Factor7	-0.09040	0.03600	-0.13151	0.07639	0.01566	-0.15213	1.00000		
Factor8	0.17493	0.24904	0.23039	0.23117	0.09145	0.14692	0.32509	1.00000	
Factor9	0.33650	-0.11167	-0.04748	0.20743	0.18806	0.00346	-0.25386	-0.22822	1.00000

Factor 1, Foundations; Factor 2, Group DSS; Factor 3, Model management;
Factor 4, User interface; Factor 5, Organizational science; Factor 6, MCDM;
Factor 7, Strategic management; Factor 8, Implementation; Factor 9 Group decision making.

RESULTS OF CLUSTER ANALYSIS AND MULTIDIMENSIONAL SCALING

Cluster analysis resulted in the Dendrogram (Figure 7.2), which illustrates hierarchical clustering of eight groups of DSS related researchers. Figure 7.2 shows both the cluster structure and the joining sequence to show how each of the authors in the study is combined into a new aggregate cluster until all 67 authors are grouped into the final one cluster (cluster one (CL1) in Figure 7.2). However, the dendrogram is constructed to show only eight clusters of DSS research. It also shows two more steps of clustering procedures below the cutoff line, if a ten or eleven cluster solution is desirable.

Cluster 8 appears to be the organizational science reference discipline, roughly corresponding to factor 5 in Table 7.1. Cluster 16 and cluster 10 combine to create cluster 8, which, in turn, is joined together with cluster 12 (user interface) to form a bigger cluster (cluster 4). The close relationship between the organizational science group and the user interface group also is represented by a high interfactor correlations coefficient (0.41) between factor 4 and factor 5. In the next clustering procedure, cluster 4 is grouped into a new bigger cluster (CL3 in Figure 7.2) with the foundations cluster (CL13). The high interfactor correlations coefficients (0.48) between the foundation factor and the organizational science factor signify the strong influence of the organizational scientists on the DSS foundations group in the development and evolution of the decision support systems field. In the next subsequent step, cluster 3 is combined with cluster 5 to form a new cluster (CL2). The dendrogram shows a linkage among various clusters, and influence could be inferred based on the close examination of the works of authors in the clusters. In this process of building cluster 2, the dendrogram shows that research in the model management area (factor 3 in Table 7.1 and cluster 14 in Figure 7.2) is linked to two reference disciplines (MCDM and organizational science) and a DSS research subspecialty (foundations). High correlations coefficients (in Figure 7.1) between the model management factor and factors 1, 5, and 6 confirm the strong

169

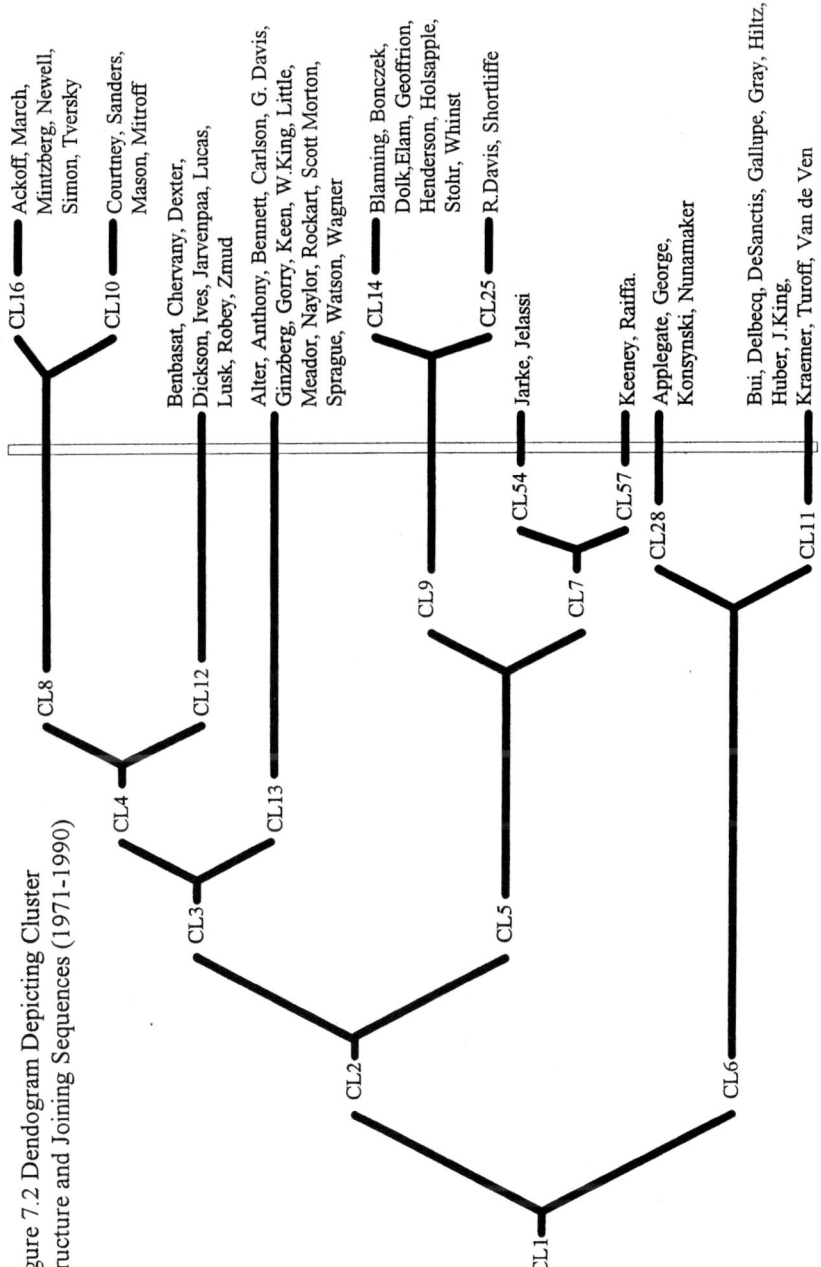

Figure 7.2 Dendogram Depicting Cluster Structure and Joining Sequences (1971-1990)

Figure 7.2 Three-Dimensional MDS Map (1970-1990)

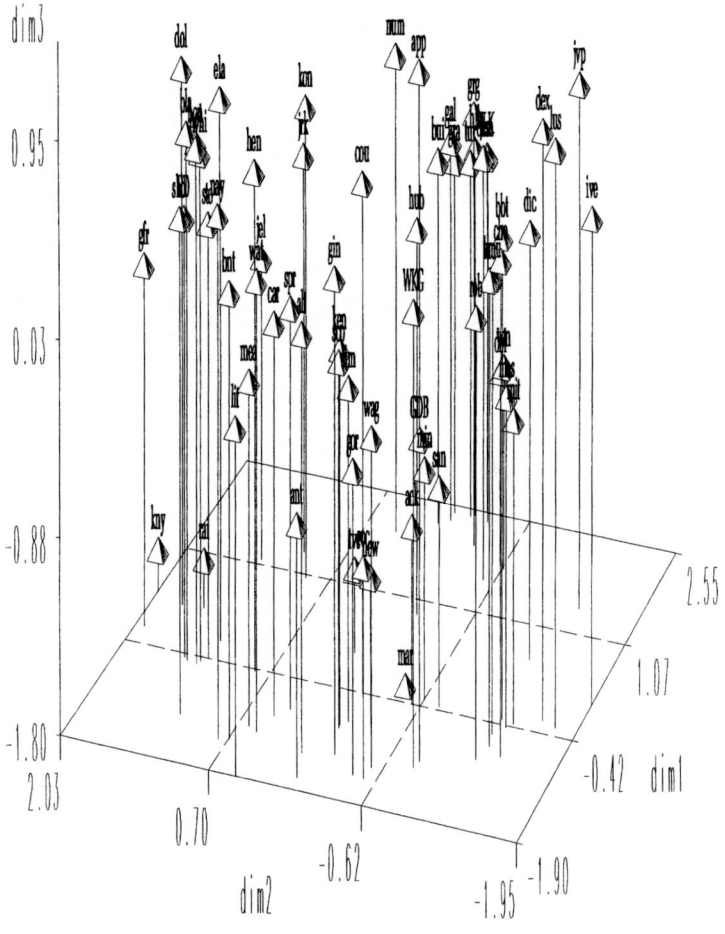

interconnections. The linkage in the dendrogram may not in and of itself demonstrate influence. In the final agglomerative clustering procedure, the GDSS and group decision making groups were the last clusters to be joined to form the final cluster (CL1), which indicates little influence of the reference disciplines except the group decision making factor.

Comparison of the two solutions from the factor analysis and cluster analysis may provide some valuable information on the similarities and differences of the two solutions to help us reach a better interpretation of the results of multivariate analysis. The two solutions reached in this study suggest that DSS research has apparently made significant progress toward a cumulative tradition in the areas of foundations, group DSS, model management, and user interface and that DSS research is grounded in the reference disciplines of organizational science and multiple criteria decision making. In addition, the last three factors (factors 7, 8, and 9) in the factor solutions must be interpreted with care. The two branches (one above and one below the cut-off line in the Dendrogram not shown in the factor analysis) may signify the role of artificial intelligence in the model management area and the emergence of the multiple criteria decision making (MCDM) model based decision support systems as an important DSS research subspecialty.

The dendrogram (Figure 7.2) clearly shows that the model management area has a linkage with cluster 52 (artificial intelligence), cluster 7 (MCDM), and cluster 3 (organizational science and other factors). Although the result of factor analysis failed to show Davis, R. and Shortliffe as an independent factor in Table 7.1, successful expert systems (ES) such as the MYCIN system (Davis et al. 1977) have been extensively examined to illustrate the concept of knowledge engineering for building business applications of ES. A recent survey of operational expert system applications in business between 1980 and 1993 revealed that few business areas remain untouched by ESs and ES-embedded systems. ESs have apparently made the transition from the research laboratory to the commercial market. ES developers have been integrating ESs with other technologies such as bar-code scanning systems, programming languages, case-based reasoning systems, natural language processing systems, robots, DSS, image processing systems, and artificial-neural-networks (Eom 1996b). These tools that combine ESs with other artificial intelligence techniques generate synergistic effects to shrink the time for tasks from days to hours, minutes, or seconds.

The three dimensional MDS map as shown in Figure 7.3 (the badness of fit value is .16) enabled us to map the overall relationships between DSS subspecialties and contributing disciplines, as well as the interrelationships within both the DSS subspecialties and the contributing disciplines to DSS. Each author's name in Figures 7.3 and 7.4 is represented by only the first three letters of each author's surname. In the case of the those authors whose last names begin with the same first three letters, such as Benbasat and Bennett, the following abbreviations are used-- BBT (Benbasat), BNT (Bennett), JLK (J.L. King), WKG (William King), GRG (George), GFR (Geoffrion), RD (Randall Davis), GDB (Gordon Davis), JRK (Jarke), JVP (Jarvenpaa), KEN (Keen), KNY (Keeney).While cluster analysis shows a detailed structure of clusters, multidimensional scaling maps show the big picture of inter-cluster relationships. The placement of authors on the center of the MDS map means that those authors are linked with a substantial portion of the author set, with relatively high correlations (e.g., founding fathers of DSS-- Keen, Scott Morton, Alter, Sprague, and Carlson). Placement near the periphery represents a more focused linkage. This is illustrated by the model management researchers (Blanning, Bonczek, Whinston, and Holsapple) located in the upper-left hand area and GDSS researchers (George, Bui, Gallupe, etc.) located in the upper-right hand area. The location of the organizational scientists, especially Tversky, Simon, Newell, and Mintzberg) in the center of the MDS map seems to indicate that organizational science is a major contributing discipline in that it connects with many other DSS research subspecialties. The strong linkage between organizational science and other DSS research subspecialties can also be supported by a major factor intercorrelation network as shown in Figure 7.1.

Figure 7.4 presents a two dimensional MDS map, which is the projection of the first and second planes presented in the three-dimensional MDS map in Figure 7.3.

Figure 7.4 Two-Dimensional MDS Map (1970-1990)

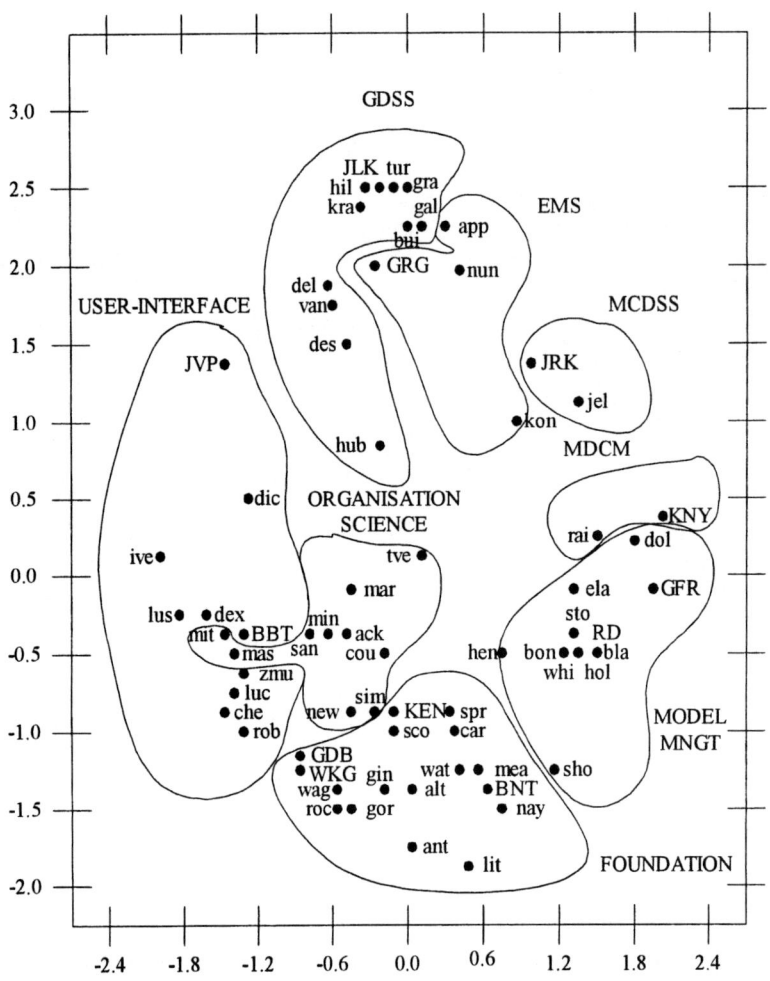

Figure 7.4 shows the organizational scientists in the center of the map, surrounded closely by the three dominant DSS research subspecialties (foundations, model management, and user interface). The GDSS cluster is somewhat isolated from the organizational science cluster and other remaining clusters. The isolated location of the GDSS cluster in Figure 7.3 and Figure 7.4 may indicate that the GDSS factor has emerged as a major DSS research subspecialty with little influence from all the other DSS contributing disciplines except for the group decision making factor.

MAJOR CONTRIBUTIONS OF ORGANIZATIONAL SCIENCE TO THE DEVELOPMENT OF DSS RESEARCH SUBSPECIALTIES

Earlier, Davis and Olson (1985) claimed that management information systems were influenced significantly by four major academic areas: managerial accounting, operations research, management and organizational theory, and computer science. Our bibliometrics-based empirical research shows the visibility of only multiple criteria decision making and organizational science to the development of DSS research subspecialties. This study, however, should not necessarily be interpreted as disproof of Davis and Olson's claim that the theory of managerial accounting was an important *MIS* reference discipline. This study shows that the *DSS* research has not been significantly influenced by the authors in management accounting, although our citing and cited database files did not exclude accounting papers. In addition to the result of this study, the result of the previous two studies (Culnan 1986; Culnan 1987) also failed to identify management accounting as a reference discipline of MIS.

Davis and Olson believe that the academic field of management information systems is an extension of management and organizational theory, which provides several important key concepts for a better understanding of the information system's functions in an organization (Davis et al. 1985, p.14). These key

concepts are as follows.
1. Behavioral theory of organizational and individual decision making.
2. Individual motivation.
3. Group process and group decision making.
4. Leadership techniques.
5. Organizational change process.
6. Organizational structure and design.

DSS are designed and implemented to support organizational as well as individual decision making. Without a detailed understanding of decision making behavior in organizations, "decision support is close to meaningless as a concept" (Keen et al. 1978, p.61). Figure 7.4 shows both the similarity and dissimilarity within each group of DSS researchers, as well as the similarity and dissimilarity among the various groups of DSS researchers. From the viewpoint of organizational science, there are three very near subspecialties--foundations, user interface/individual differences, and model management. The MCDSS subspecialty appears to be the next research subspecialty which has been influenced by organizational scientists. Figure 7.1 shows the interfactor correlations among 9 factors. The organizational science factor has high correlation coefficients with factor 1 (foundations), factor 3 (model management), and factor 4 (user interface).

CONTRIBUTIONS OF ORGANIZATIONAL SCIENCE TO THE FOUNDATIONS OF DECISION SUPPORT SYSTEMS

As Huber (1981) pointed out, organizational decision making is concerned with how one or more organizational units make a decision on behalf of the organization. Organizational scientists classified organizational decision making in terms of several schools of thought: 1) the rational model, as depicted by such scholars as Mintzberg (1973), and Mintzberg, Raisinghani, and Theoret (1976), which focuses on the selection of the most efficient alternatives, with the assumption of a rational, completely informed, single decision maker; 2) the organizational process model developed by Cyert and March (1963), which stresses the compartmentalization of the various units in any organization; 3) the satisficing model by Simon and his colleagues (Newell et al. 1972), which illustrates finding an acceptable, "good enough" solution, and thus reflects "bounded rationality"; and 4) the garbage can model, which is defined by Huber (1981, p.48) as: "Organizational decisions are consequences of intersections of problems looking for solutions, solutions looking for problems, and opportunities for making decisions." Huber's study (1981) was the first one that addressed a part of our concern. He concluded that there exist the following three organizational science contributions to the design of DSS. First, "DSSs that are compatible with the manager's dss will be more used and useful than those that are not" (Huber 1981, p.46). He distinguished a DSS from a dss. He defined the term "dss" as "the information sources and decision-aiding processes that he or she draws upon as the occasion requires." Second, organizational scientists contributed to the design of DSS by modeling the empirical and normative structures and dynamics of organizational information systems. Third, organizational scientists contributed to the design of DSS by modeling the nature of managerial decision making processes in general.

Simon's *New Science of Management Decisions* (Simon 1960) has been one of the widely-cited books in the DSS area. There are several foundational concepts that supported the emergence of decision support systems, including:

1. three principal stages of decision making (intelligence, design, and choice), and
2. the distinction between programmed and non-programmed decisions.

His foresight resulted in a critical contribution to the development of DSS. Computers should be used to deal with non-programmable decisions by integrating management science/operations research models into the computer systems for managerial decision making and by using symbol manipulation capabilities. In the early 1960s, Simon (1960) posited that computers could be used as a tool for simulating human thought through the manipulation of words (symbols).

Simon's *Administrative Behavior* (Simon 1976) viewed organizations in terms of their decision process:

> *Administrative Behavior* has served me as a useful and reliable port of embarkation for voyages of discovery into human decision making: the relation of organization structure to decision making, the formalized decision making of operations research and management science, and in more recent years, the thinking and problem-solving activities of individual human beings. (ix)

Mintzberg's model has been especially useful in conceptualizing the relationship between information flows and managerial roles, as a basis for information system design. An important contribution of Mintzberg is his clear definition of ten interrelated roles that managers perform to achieve the basic organizational purpose. These ten roles are categorized into three major groupings: interpersonal roles (which derive from the manager's authority and status), informational roles (which derive from the interpersonal roles and the access they provide to information), and decisional roles (which derive from the manager's authority and information) (Mintzberg 1973).

Bariff and Ginzberg (1982) had earlier presented a framework for classifying and analyzing behavioral information systems research and depicted ties to behavioral science disciplines. They further identified major research issues in the design and implementation of MIS, use of MIS, and management of MIS functions at each of the four different levels (individual, group, organizational, interorganizational).

CONTRIBUTIONS OF ORGANIZATIONAL SCIENCE TO THE USER INTERFACE/INDIVIDUAL DIFFERENCES

The initial investigation of this topic was begun by Mason and Mitroff (1973, p.478), in whose earlier works they hypothesized that "the designers of information systems should not force all psychological types to conform to one type of information system, rather each psychological type should be given the kind of information to which he is psychologically attuned and uses most effectively."

Newell and Simon's *Human Problem Solving* included the study of the individual difference dimension of human behavior. "Each man differs--both in systematic ways and simply by virtue of his unique genetic endowment and historical fate--from all other men" (Newell et al. 1972, p.3). Simon's work (Simon 1976) also played an important role in the development of the individual differences research subspecialty in the DSS area. He pointed out that for the individual to be equipped to make the correct decisions, the organization must place that individual in a psychological environment that will adapt his or her decisions to the organization's objectives and that will provide the information needed to make decisions correctly.

CONTRIBUTIONS OF ORGANIZATIONAL SCIENCE TO MODEL MANAGEMENT

Figure 7.1 shows a high correlation between the organizational science factor and the model management factor, which probably may have resulted from the nine author overlap between those loading above ± 0.4.

As discussed in the earlier section, the central theme of Simon's work is the theory of intentional and bounded rationality-- the behavior of human beings who satisfice because they are unable to maximize (Simon 1976, p. xxviii). Simon's theory of bounded rationality has served as a foundational concept for multiple criteria decision making, which, in turn, opened a new line of inquiry in the model management area-- Should Model management in single-user DSS be different from the one for multiperson-user DSS?

CONCLUSIONS

We have identified several decision support systems research subspecialties based upon factor analysis, multidimensional scaling, and cluster analysis of a massive amount of bibliographical data. Two major conclusions from this study are that decision support systems are *weakly* grounded in four contributing disciplines (organizational science, multiple criteria decision making, group decision making, and strategic management) and that researchers in the DSS research area are in the process of developing their own theories in the areas of foundations, GDSS, model management, user interface/individual differences, and DSS implementation. This research provides a piece of evidence of the maturing process and the existence of the cumulative research tradition in the DSS area.

Organizational scientists have made important contributions to the development of such DSS research subspecialties as foundations, user interface/individual differences, and model management. An important implication of this result is that although DSS writers have referred to such parent disciplines as psychology, political science, accounting, computer science, and management science/operations research, these disciplines were not visible in this study and have apparently not played as important a role in the evolution of current DSS research as some previous commentators have opined.

Over the last two decades (1970-1990), DSS research has mainly concentrated on each component (data/model/dialogue/decision maker) of specific DSS. Keen and Scott Morton's DSS frameworks based on organizational perspectives have not been widely adopted over the past two decades. Our results show that none of Keen's suggested areas of DSS study (design, implementation, evaluation of DSS) have been shown to be substantive DSS research subspecialties. Numerous, dedicated researchers in the DSS implementation area have attempted to systematically identify the implementation success factors and the relationship between user-related factors (cognitive style, personality, demographics, and user-situational variables) and implementation success.

Nevertheless, Alavi and Joachimsthaler (1992, p. 95) concluded that "Although information systems implementation has been a topic of interest to researchers over the past two decades, the extent to which the existing body of research reflects substantial and cumulative development is not entirely clear" based on a meta-analysis of 144 findings. Future DSS research should redirect its attention to these undernourished areas of DSS research to provide useful guiding principles for practitioners in the integrated process of design, implementation, and evaluation.

The accelerating rate of DSS research publication and steady transition from non-empirical to empirical studies as shown in this research have not resulted in DSS theory building. This research has provided hard evidence that most empirical DSS research areas (e.g., GDSS, Individual differences, and DSS implementation) have produced an accumulation of conflicting results due to methodological problems, lack of a commonly accepted causal model, different measures of dependent variables, hardware and software designed under different philosophies, different tasks. Nevertheless, encouraging developments are being made to integrate the seemingly conflicting results of numerous GDSS empirical experiments and to integrate inconsistent results on how humans process information presented in graphical form into useful frameworks for better understanding different forms of graphical representation.

This research identified a group of influential and responsible DSS researchers, representing the major forces that will chart the future directions for DSS research and redirect the DSS research efforts to become paradigm-directed. "Any study of paradigm-directed or of paradigm-shattering research must begin by locating the responsible group or groups" (Kuhn 1970, p. 242.).

REFERENCES

Ackoff, R.L. "Management Misinformation Systems," *Management Science* (14:12), December 1967, pp B147-B156.

Alavi, M., and Joachimsthaler, E.A. "Revisiting DSS Implementation Research: A Meta-Analysis of the Literature and Suggestions for Researchers," *MIS Quarterly* (16:1), March 1992, pp 95-116.

Alter, S.L. "A Study of Computer Aided Decision Making in Organizations," MIT, Boston, MA, 1975.

Alter, S.L. "A Taxonomy of Decision Support Systems," *Sloan Management Review* (19:1), Fall 1977, pp 39-56.

Alter, S.L. *Decision Support Systems: Current Practice and Continuing Challenges*, Addison-Wesley, Reading, MA, 1980.

Anthony, R.N. *Planning and Control Systems: A Framework for Analysis*, Division of Research, Graduate School of Business, Harvard University, Cambridge, MA, 1965.

Baker, D.R. "Citation Analysis: A Methodological Review," *Social Work Research & Abstracts* (26:3), September 1990, pp 3-10.

Bariff, M.L., and Ginzberg, M.J. "MIS and Behavioral Sciences: Research Patterns and Prescriptions," *Data Base* (14:1) 1982, pp 19-26.

Bariff, M.L., and Lusk, E.J. "Cognitive and Personality Tests for the Design of Management Information Systems," *Management Science* (23:8), April 1977, pp 820-829.

Bayer, A.E., Smart, J.C., and McLaughlin, G.W. "Mapping Intellectual Structure of a Scientific Subfield through Author Cocitations," *Journal of the American Society for Information Science* (41:6), September 1990, pp 444-452.

Bellardo, T. "The Use of Co-Citations to Study Science," *Library Research* (2) 1980, pp 231-237.

Benbasat, I. "An Experimental Evaluation of the Effects of Information System and Decision Maker Characteristics on Decision Effectiveness," in: *Department of Information and Decision Sciences*, University of Minnesota, Minneapolis, MN, 1974.

Benbasat, I., DeSanctis, G., and Nault, B.R. "Empirical Research in Managerial Support Systems: A Review and Assessment," in: *Recent Development in Decision Support Systems*, C.W. Holsapple and A.B. Whinston (eds.), Springer-Verlag, Berlin, 1993, pp. 383-437.

Benbasat, I., and Dexter, A.S. "Value and Events Approaches to Accounting: An Experimental Evaluation," *Accounting Review* (54:4), October 1979, pp 735-749.

Benbasat, I., and Dexter, A.S. "Individual Differences in the Use of Decision Support Aids," *Journal of Accounting Research* (20:1), Spring 1982, pp 1-

11.
Blanning, R.W. "A Relational Framework for Model Management in Decision Support Systems," in: *Decision Support Systems-82 Transactions*, G.W. Dickson (ed.), San Francisco, CA, 1982, pp. 16-28.

Blanning, R.W. "Model Management Systems: An Overview," *Decision Support Systems* (9:1), January 1993, pp 9-18.

Bonczek, R.H., Holsapple, C.W., and Whinston, A.B. "Computer-Based Support of Organizational Decision Making," *Decision Sciences* (10:2), April 1979, pp 268-291.

Bonczek, R.H., Holsapple, C.W., and Whinston, A.B. "The Evolving Roles of Models in Decision Support Systems," *Decision Sciences* (11:2), April 1980a, pp 337-356.

Bonczek, R.H., Holsapple, C.W., and Whinston, A.B. "Future Directions for Developing Decision Support Systems," *Decision Sciences* (11:4), October 1980b, pp 616-631.

Bonczek, R.H., Holsapple, C.W., and Whinston, A.B. *Foundations of Decision Support Systems*, Academic Press, New York, 1981.

Bui, T.X., and Jarke, M.A. "A DSS for Cooperative Multiple Criteria Group Decision Making," in: *Proceedings of the 5th International Conference on Information Systems*, Tucson, AZ, 1984, pp. 101-113.

Cattell, R.B. "The Scree Test for the Number of Factors," *Multivariate Behavioral Research* (1), April 1966, pp 245-276.

Chang, A.-M., Holsapple, C.W., and Whinston, A.B. "Model Management Issues and Directions," *Decision Support Systems* (9:1), January 1993, pp 19-37.

Chervany, N.L., and Dickson, G.W. "An Experimental Evaluation of Information Overload in a Production Environment," *Management Science* (20:10), June 1974, pp 1335-1344.

Child, D. *The Essentials of Factor Analysis*, Holt, Rinehart and Winston, Inc., London, 1970.

Cooper, D.R., and Emory, C.W. *Business Research Methods*, (Fifth ed.), Irwin, Chicago, IL, 1995.

Crane, D. *Invisible Colleges: Diffusion of Knowledge in Scientific Communities*, University of Chicago Press, Chicago, 1972.

Culnan, M.J. "The Intellectual Development of Management Information Systems, 1972-1982: A Co-Citation Analysis," *Management Science* (32:2), February 1986, pp 156-172.

Culnan, M.J. "Mapping the Intellectual Structure of MIS, 1980-1985: A Co-Citation Analysis," *MIS Quarterly* (11:3), September 1987, pp 341-353.

Cushing, B.E. "Frameworks, Paradigms, and Scientific Research in Management Information Systems," *Journal of Information Systems* (4:2), Spring 1990, pp 38-59.

Cyert, R.M., and March, J.G. *A Behavioral Theory of the Firm*, Prentice Hall Behavioral Sciences in Business Series, Prentice Hall, Englewood Cliffs, NJ, 1963.

Davis, G.B., and Olson, M.H. *Management Information Systems: Conceptual*

Foundations, Structure and Development, (2nd ed.), McGraw-Hill, New York, 1985.

Davis, R., Buchanan, B.G., and Shortliffe, E. "Production Rules as a Representation for a Knowledge-Based Consultation Program," *Artificial Intelligence* (8:1) 1977, pp 15-45.

Delbecq, A.L., Van de Ven, A.H., and Gustafson, D.H. *Group Techniques for Program Planning: A Guide to Nominal Group and Delphi Processes*, Scott, Foresman and Company, Glenview, IL, 1975.

Dennis, A.R., and Gallupe, R.B. "A History of Group Decision Support Systems Empirical Research: Lessons Learned and Future Directions," in: *Group Support Systems: New Perspectives*, L.M. Jessup and J.S. Valacich (eds.), Macmillan, New York, 1993, pp. 59-77.

Dennis, A.R., George, J.F., Jessup, L.M., Nunamaker, J.F., Jr., and Vogel, D.R. "Information Technology to Support Electronic Meetings," *MIS Quarterly* (12:4), December 1988, pp 591-624.

DeSanctis, G. "Computer Graphics as Decision Aids: Directions for Research," *Decision Sciences* (15:4), Fall 1984, pp 463-487.

DeSanctis, G., and Gallupe, B. "A Foundation for the Study of Group Decision Support Systems," *Management Science* (33:5), May 1987, pp 589-609.

Dickson, G.W., Senn, J.A., and Chervany, N.L. "Research in Management Information Systems: The Minnesota Experiments," *Management Science* (23:9), May 1977, pp 913-923.

Dolk, D.R., and Konsynski, B.R. "Knowledge Representation for Model Management Systems," *IEEE Transactions On Software Engineering* (SE-10:6), November 1984, pp 619-628.

Dolk, D.R., and Kottemann, J.E. "Model Integration and a Theory of Models," *Decision Support Systems* (9:1), January 1993, pp 51-63.

Dyer, J.S., Fishburn, P.C., Steuer, R.E., Wallenius, J., and Zionts, S. "Multiple Criteria Decision Making, Multiattribute Utility Theory: The Next Ten Years," *Management Science* (38:5), May 1992, pp 645-654.

Elam, J.J., Henderson, J.C., and Miller, L.W. "Model Management Systems: An Approach to Decision Support in Complex Organizations," in: *Proceedings of the First International Conference on Information Systems*, E.R. McLean (ed.), Philadelphia, PA, 1980, pp. 98-110.

Elam, J.J., Huber, G.P., and Hurt, M.E. "An Examination of the DSS Literature (1975-1985)," in: *Decision Support Systems: A Decade in Perspective*, E.R. McLean and H.G. Sol (eds.), Elsevier Science, Amsterdam, North-Holland, 1986, pp. 239-251.

Eom, H.B., and Lee, S.M. "Decision Support Systems Applications Research: A Bibliography (1971-1988)," *European Journal of Operational Research* (46:3), June 15 1990a, pp 333-342.

Eom, H.B., and Lee, S.M. "A Survey of Decision Support System Applications (1971-April 1988)," *Interfaces* (20:3), May-June 1990b, pp 65-79.

Eom, S.B. "Ranking Institutional Contributions to Decision Support Systems Research: A Citation Analysis," *Data Base* (25:1), February 1994, pp 35-

42.

Eom, S.B. "Decision Support Systems Research: Reference Disciplines and a Cumulative Tradition," *Omega: The International Journal of Management Science* (23:5), October 1995, pp 511-523.

Eom, S.B. "Mapping the Intellectual Structure of Research in Decision Support Systems through Author Cocitation Analysis (1971-1993)," *Decision Support Systems* (16:4), April 1996a, pp 315-338.

Eom, S.B. "A Survey of Operational Expert Systems in Business (1980-1993)," *Interfaces* (26:5), September-October 1996b, pp 50-70.

Eom, S.B. "Assessing the Current State of Intellectual Relationships between the Decision Support Systems Area and Academic Disciplines," The Eighteenth International Conference on Information Systems, Atlanta, GA, 1997, pp. 167-182.

Eom, S.B. "The Intellectual Development and Structure of Decision Support Systems (1991-1995)," *Omega* (26:5), October 1998a, pp 639-658.

Eom, S.B. "Relationships between the Decision Support System Subspecialties and Reference Disciplines: An Empirical Investigation," *European Journal of Operational Research* (104:1), 1-Jan 1998b, pp 31-45.

Eom, S.B. *Decision Support Systems Research (1970-1999): A Cumulative Tradition and Reference Disciplines*, Edwin Mellen Press, Lewiston, New York, 2002, p. 412.

Eom, S.B., and Farris, R. "The Contributions of Organizational Science to the Development of Decision Support Systems Research Subspecialties," *Journal of the American Society for Information Science* (47:12), December 1996, pp 941-952.

Eom, S.B., and Lee, S.M. "Leading Universities and Most Influential Contributors in DSS Research: A Citation Analysis," *Decision Support Systems* (9:3), April 1993a, pp 237-244.

Eom, S.B., Lee, S.M., and Kim, J.K. "The Intellectual Structure of Decision Support Systems (1971-1989)," *Decision Support Systems* (10:1), July 1993b, pp 19-35.

Everitt, B.S. *Cluster Analysis*, Heinemann Educational Books Ltd., London, 1980.

Farhoomand, A.F. "Scientific Progress of Management Information Systems," *Data Base* (18:4), Summer 1987, pp 48-56.

Gallupe, R.B., DeSanctis, G., and Dickson, G.W. "The Impact of Computer Support on Group Problem Finding: An Experimental Approach," *MIS Quarterly* (12:2), June 1988, pp 276-296.

Garfield, E. "Citation Indexes for Science," *Science* (122) 1955, pp 108-111.

Garfield, E. "Can Citation Indexing Be Automated?," in: *Statistical Association Methods for Mechanized Documentation (Nbs Misc. Pub. 269.)*, M.E. Stevens, et al. (ed.), National Bureau of Standards, Washington, D.C., 1965, p. 189.

Garfield, E. *Citation Indexing: Its Theory and Application in Science, Technology, and Humanities*, Wiley, New York, 1979.

Geoffrion, A.M. "An Introduction to Structured Modeling," *Management Science*

(33:5), May 1987, pp 547-588.
Geoffrion, A.M., Dyer, J.S., and Feinberg, A. "An Interactive Approach for Multicriteria Optimization with an Application to the Operation of an Academic Department," *Management Science* (19:4) 1972, pp 357-368.
Ginzberg, M.J. "Early Diagnosis of MIS Implementation Failure: Promising Results and Unanswered Questions," *Management Science* (27:4), April 1981, pp 459-478.
Gorry, G.A., and Scott Morton, M.S. "A Framework for Management Information Systems," *Sloan Management Review* (13:1), Fall 1971, pp 55-70.
Hair, J.F., Jr., Anderson, R.E., Tatham, R., and Black, W.C. *Multivariate Data Analysis with Readings*, (3rd ed. ed.), Macmillan Publishing, New York, 1992.
Hair, J.F., Jr., Anderson, R.E., and Tatham, R.L. *Multivariate Data Analysis with Readings*, (2nd ed.), Macmillan Publishing Company, New York, 1987.
Hatcher, L. *A Step-by-Step Approach to Using the SAS System for Factor Analysis and Structural Equation Modeling*, (3rd ed. Printing ed.), SAS Institute Inc., Carry, NC, 1994, p. 588.
Hoffman, D.L., and Holbrook, M.B. "The Intellectual Structure of Consumer Research: A Bibliometric Study of Author Cocitations in the First 15 Years of the Journal of Consumer Research," *Journal of Consumer Research* (19), March 1993, pp 505-517.
Huber, G.P. "The Nature of Organizational Decision Making and the Design of Decision Support Systems," *MIS Quarterly* (5:2), June 1981, pp 1-10.
Huber, G.P. "Cognitive Style as a Basis for MIS and DSS Design: Much Ado About Nothing?," *Management Science* (29:5), May 1983, pp 567-579.
Huber, G.P. "Issues in the Design of Group Decision Support Systems," *MIS Quarterly* (8:3), September 1984, pp 195-204.
Hulme, E.W. *Statistical Bibliography in Relation to the Growth of Modern Civilization*, Grafton, London, 1923.
Ives, B. "Graphical User Interfaces for Business Information Systems," *MIS Quarterly* (6:4), December 1982, pp 15-46.
Jarvenpaa, S.L., Dickson, G.W., and DeSanctis, G. "Methodological Issues in Experimental Is Research: Experiences and Recommendations," *MIS Quarterly* (9:2), June 1985, pp 141-156.
Jelassi, M.T., Jarke, M.A., and Stohr, E.A. "Designing a Generalized Multiple Criteria Decision Support System," *Journal of Management Information Systems* (1:4), Spring 1985, pp 24-43.
Keen, P.G.W. "MIS Research: Reference Disciplines and a Cumulative Tradition," in: *Proceedings of the First International Conference on Information Systems*, E.R. McLean (ed.), Philadelphia, PA, 1980, pp. 9-18.
Keen, P.G.W., and Scott Morton, M.S. *Decision Support Systems: An Organizational Perspective*, Addison-Wesley, Reading, MA, 1978.
Keeney, R.L., and Raiffa, H. *Decisions with Multiple Objectives: Preferences and Value Tradeoffs*, John Wiley and Sons, New York, 1976.

Khattree, R., and Naik, D.N. *Applied Multivariate Statistics with SAS Software*, (2nd ed.), SAS Institute, Cary, NC, 1999.

Kim, J., and Mueller, C.W. *Factor Analysis: Statistical Methods and Practical Issues*, Sage Publications, Inc., Beverly Hills, CA, 1978.

King, W.R., and Rodriguez, J.I. "Evaluating Management Information Systems," *MIS Quarterly* (2:3), September 1978, pp 43-51.

Konsynski, B.R., Kottemann, J.E., Nunamaker, J.F., Jr., and Stott, J.W. "Plexsys-84: An Integrated Development Environment for Information System," *Journal of Management Information Systems* (1:3), Winter 1984-1985, pp 64-104.

Kraemer, K.L., and King, J.L. "Computer-Based Systems for Cooperative Work and Group Decision Making," *ACM Computing Surveys* (20:2), June 1988, pp 115-146.

Kruskal, J.B., and Wish, M. *Multidimensional Scaling*, Sage University Paper Series on Quantitative Applications in the Social Sciences, Sage Publications, Beverly Hills, CA, 1978.

Kruskal, J.B., and Wish, M. *Multidimensional Scaling*, Sage University Paper Series on Quantitative Applications in the Social Sciences, 07-011. Sage Publications, Beverly Hills and London, 1990.

Kuhn, T.S. *The Structure of Scientific Revolutions*, (2d ed. ed.), The University of Chicago Press, Chicago, Ill., 1970.

Kuo, A. *The Distance Macro: Preliminary Documentation*, (2nd ed.), The SAS Institute, Cary, NC, 1997, p. 31.

Lindsey, D. "Production and Citation Measures in the Sociology of Science: The Problem of Multiple Authorship," *Social Studies of Science* (10) 1980, pp 145-162.

Little, J.D.C. "Models and Managers: The Concepts of a Decision Calculus," *Management Science* (16:8), April 1970, pp B466-B485.

Long, J.S., et al. "The Problem of Junior-Authored Papers in Constructing Citation Counts," *Social Studies of Science* (10), May 1980, pp 127-143.

Lucas, H.C., Jr. "An Experimental Investigation of the Use of Computer-Based Graphics in Decision Making," *Management Science* (27:7), July 1981, pp 757-768.

Lucas, H.C., Jr., and Nielsen, N.R. "The Impact of the Mode of Information Presentation on Learning and Performance," *Management Science* (26:10), October 1980, pp 982-993.

Lusk, E.J., and Kersnick, M. "The Effect of Cognitive Style and Report Format on Task Performance: The MIS Design Consequences," *Management Science* (25:8), August 1979, pp 787-798.

Mason, R.O., and Mitroff, I.I. "A Program for Research on Management Information Systems," *Management Science* (19:5), January 1973, pp 475-487.

McCain, K.W. "Longitudinal Author Cocitation Mapping: The Changing Structure of Macroeconomics," *Journal of the American Society for Information Science* (35:6), January 1984, pp 351-369.

McCain, K.W. "Cocited Author Mapping as a Valid Representation of Intellectual Structure," *Journal of the American Society for Information Science* (37:3) 1986, pp 111-122.

McCain, K.W. "Mapping Authors in Intellectual Space: A Technical Overview," *Journal of the American Society for Information Science* (41:6), September 1990, pp 351-359.

Mintzberg, H. *The Nature of Managerial Work*, Prentice Hall, Englewood Cliffs, NJ, 1973.

Mintzberg, H., Raisinghani, D., and Theoret, A. "The Structure of "Unstructured" Decision Processes," *Administrative Science Quarterly* (21:2), June 1976, pp 246-275.

Newell, A., and Simon, H.A. *Human Problem Solving*, Prentice Hall, Englewood Cliffs, NJ, 1972.

Nunamaker, J.F., Jr., Applegate, L.M., and Konsynski, B.R. "Facilitating Group Creativity: Experience with a Group Decision Support System," *Journal of Management Information Systems* (3:4), Spring 1987, pp 5-19.

Pinsonneault, A., and Kraemer, K.L. "The Impact of Technological Support on Groups: An Assessment of the Empirical Research," *Decision Support Systems* (5:2) 1989, pp 197-216.

Prichard, A. "Statistical Bibliography or Bibliometrics?," *Journal of Documentation* (25:4), December 1969, pp 348-349.

Raisig, L.M. "Statistical Bibliography in the Health Science," *Bulletin of Medical Library Association* (50:3), July 1962, pp 45-461.

Ramsay, J.O. *The Mlscale Procedure*, Sugi Supplementary Library User's Guide (Version 5 Ed.), SAS Institute, Cary, NC, 1986.

Rosengren, K.E. *Sociological Aspects of the Literary System*, Natur och Kultur, Stockholm, 1968.

Rosengren, K.E. "Who Carries the Field? Communications between Literary Scholars and Critics," in: *Scholarly Communication and Bibliometrics*, C.L. Borgman (ed.), Sage, Newbury Park, CA, 1990, pp. 107-128.

Sanders, G.L., and Courtney, J.F. "A Field Study of Organizational Factors Influencing DSS Success," *MIS Quarterly* (9:1), March 1985, pp 77-93.

SAS Institute Inc. *SAS/Stat User's Guide, Release 6.03*, (3rd ed. ed.), SAS Institute Inc., Cary, NC., 1988, pp. 449-466.

SAS Institute Inc. *SAS Technical Report P-229, SAS/Stat Software: Changes and Enhancements (Release 6.07)*, SAS Institute Inc., Cary, NC, 1992, pp. 251-286.

SAS Institute Inc. *SAS Onlinedoc® Version 8, February 2000, Pdf Format*, SAS Institute Inc., Cary, NC, 2000.

Simon, H.A. *The New Science of Management Decision*, Harper & Row, New York, 1960.

Simon, H.A. *The Sciences of the Artificial*, (2nd ed. ed.), The MIT Press, Cambridge, MA, 1969.

Simon, H.A. *Administrative Behaviour: A Study of Decision Making Processes in Administrative Organization*, The Free Press, A Division of Macmillan

Publishing Co., New York, 1976.

Smith, L.C. "Citation Analysis," *Library Trends* (30:1), Summer 1981, pp 83-106.

Sprague, R.H., Jr. "A Framework for the Development of Decision Support Systems," *MIS Quarterly* (4:4) 1980, pp 1-26.

Sprague, R.H., Jr. , and Carlson, E.D. *Building Effective Decision Support Systems*, Prentice Hall, Englewood Cliffs, NJ, 1982.

Sprague, R.H., Jr. , and Watson, H.J. "MIS Concepts: Part II," *Journal of Systems Management* (26:2), February 1975, pp 35-40.

Sprague, R.H., Jr., and Watson, H.J. "DSS Bibliography," in: *Decision Support Systems: Putting Theory into Practice*, R.H. Sprague, Jr. and H.J. Watson (eds.), Prentice Hall, Englewood Cliffs, NJ, 1989, pp. 403-413.

Tan, J.K.H., and Benbasat, I. "The Effectiveness of Graphical Presentation for Information Extraction: A Cumulative Experimental Approach," *Decision Sciences* (24:1), January-February 1993, pp 167-191.

Teng, J.T.C., and Galletta, D.F. "MIS Research Directions: A Survey of Researcher's Views," *Data Base* (21:3/4), Fall 1990, pp 1-10.

Turoff, M., and Hiltz, S.R. "Computer Support for Group Versus Individual Decisions," *IEEE Transactions On Communications* (COM-30:1), January 1982, pp 82-92.

Wagner, G.R. "Decision Support Systems: The Real Substance," *Interfaces* (11:2), April 1981, pp 77-86.

White, H.D. "Cocited Author Retrieval Online: An Experiment with the Social Indicators Literature," *Journal of American Society for Information Science* (32) 1981, pp 16-22.

White, H.D. "A Cocitation Map of the Social Indicators Movement," *Journal of the American Society for Information Science* (34:5) 1983, pp 307-312.

White, H.D., and Griffith, B.C. "Author Cocitation: A Literature Measure of Intellectual Structure," *Journal of the American Society for Information Science* (32:3) 1981, pp 163-171.

White, H.D., and Griffith, B.C. "Authors as Markers of Intellectual Space: Cocitation in Studies of Science, Technology, and Society," *Journal of Documentation* (38) 1982, pp 255-272.

Young, F.W., and Hamer, R.M. *Multidimensional Scaling: History, Theory, and Applications*, Lawrence Erlbaum Associates, Publishers, Hillsdale, New Jersey, 1987.

Young, F.W., Lewyckyj, R., and Takane, Y. *The Alscal Procedure*, Sugi Supplementary Library User's Guide (Version 5 Edition), SAS Institute, Cary, NC, 1986.

SUBJECT INDEX

%

%Distance Arguments, Ii, 81
%INCLUDE, Ii, 78, 92, 154

A

Administrative Behavior, 178
Agglomerative Procedure, 88
Annotate Data Set, Iii, I, 131, 132, 133
Artificial Intelligence (Ai), 146, 157, 162, 163, 171, 172
Aspect Ratio, 115
Assignment Statements, 136
Author Cocitation Analysis (Aca), 7, 8, 9, 11, 12, 23, 28, 35, 36, 40, 41, 54, 88, 150
Author Cocitation Matrix, 93, 101, 148, 152

B

Bar-Code Scanning Systems, 172
Bibliographic Databases, I, Ii, 3, 4, 6, 11, 13, 15, 24, 26, 29, 68, 148
Bibliometrics, 7, 8, 9, 175
Bounded Rationality, 157, 177, 179
Box, I, 116, 117, 118, 126, 127, 140, 141

C

Cards, 49, 85
Cards Statement, 48
Case-Based Reasoning, 172
Citation Analysis, 12
Cluster Analysis, 42, 77, 85, 87, 93, 98, 100, 122, 145, 150, 167, 171, 172, 180
Cluster History, Iii, 43, 89, 91, 94, 95
Cluster Procedure, 84
Clustering Algorithms, 87, 88
Cocitaion Counts, 29
Cocitation Analysis, 8, 9, 11, 150, 158
Cocitation Frequency, 11, 24, 35, 103, 145, 150
Cocitation Frequency Matrix, 24, 103, 145, 150
Cocitation Matrix Generation System, Iv, 28, 35, 150, 151
Cognitive Style, 164, 180
Common Factor Analysis, 54
Communality, I, 64, 66, 72
Component Analysis, 54
Computer Science, 157, 175, 180
Contributing Disciplines, 12, 23, 45, 146, 157, 172, 175, 180
Conventional Interacting (Discussion) Group, 158, 160
Correlation Matrix, 54, 152, 153
Cumulative Research Tradition, 145, 146, 147, 148, 158, 171, 180

D

DATA Statement, 48, 85
Database Management Systems (DBMS), 146, 148
DATALINES Statement, 49
Decision Calculus, 159
Decision Room, 161
Decision Support Systems (DSS), 23, 87, 145, 146, 147, 148, 158, 159, 160, 162, 167, 171, 177, 180
Architecture, 159

DSS, 19, 23, 27, 28, 45, 60, 98, 101, 145, 146, 147, 148, 149, 150, 151, 152, 153, 157, 158, 159, 160, 162, 164, 165, 167, 171, 172, 175, 176, 177, 179, 180, 181
Implementation, 160, 165, 180, 181
Research Subspecialties, 29, 45, 145, 147, 151, 153, 157, 172, 175, 180
Single User, 161
Specific, 146, 148, 149, 180
Decisional Roles, 178
Delphi Technique, 160
Dendrogram, I, Ii, 98, 99, 167, 169, 171
Dependent Variable, 41, 45
Dimension Coefficients, Iii, 106, 109
DISTANCE Macro, 77, 78, 152
Distance Measures, Ii, 78, 82, 88
Distnew.Sas, 78, 85, 86, 92, 101, 125, 139, 141, 154
Distnew.Sas, Ii, 78
Document Cocitation Analysis, 9

E

Eigenvalue, 55
Electronic Meeting Systems (EMS), 162
Entity-Relationship Data Model, 162
Euclidian Distance, 83
Expert Systems (ES), 171

F

Factor Analysis, 42, 45, 87, 100, 145, 150, 153, 171, 180
Factor Extraction Methods, 54
Factor Loadings, 27, 56, 63, 152
Factor Structure Correlations, I, 75
Fox-Base Based Matrix Generation System, 150
Function Statement, 136

G

G3D Procedure, 128
G3D Procedures, 6, 101
Garbage Can Model, 177
Group Decision Making, 153, 158, 160, 171, 175, 176, 180
Group Decision Support System (GDSS), 145, 146, 148, 153, 160, 161, 171, 172, 175, 180, 181

H

HAXIS, I, 116, 117, 118, 119, 126, 127, 140, 141, 142
Hierarchical Clustering, 98, 167

I

ID Statement, Iii, 81, 89, 108
Image Processing Systems, 172
Implementation Success Factors, 180
INFILE Statement, Ii, 51, 59
Information Systems, 146, 147, 148, 159, 164, 175, 177, 178, 181
Informational Roles, 178
INPUT Statement, 48, 49, 85
Institute For Scientific Information (ISI), 30, 150
Intellectual Structure, 9, 11, 12, 23, 27, 55, 67, 145, 148, 149, 150
Interfactor Correlations, 73, 167, 176
Inter-Object Similarity, 78, 87, 102

K

Knowledge-Based Model Management Systems, 163

L

Latent Root Criterion (Eigenvalue 1 Criterion), 63, 152
LENGTH Statement, 134
Level-Of-Adoption Hypothesis, 160

Libname, 53, 59, 86, 90, 105, 113, 127, 128, 129, 131, 133, 141
Local Area Decision Nets (LADNS), 161

M

Management Accounting, 175
Management Information Systems (MIS), 12, 13, 146, 147, 148, 158, 159, 160, 163, 175, 178
Management Science/Operations Research (MS/OR), 175, 177, 178, 180
Management Support Systems (MSS), 158
Managerial Accounting, 175
Marketing DSS, 146
Mean Cocitation Rate, 28, 150, 151
Metric Variables, 42
MINEIGEN, Ii, 52, 55, 58, 59, 64, 66
Model Base Processing, 162
Model Integration, 162
Model Management, 146, 148, 153, 162, 163, 167, 171, 172, 175, 176, 179, 180
Model Management Systems (MMS), 148, 162, 163
Model/Data Management, 145, 162
Multidimensional Scaling (MDS), 84, 87, 93, 101, 102, 103, 105, 106, 107, 113, 122, 145, 148, 150, 152, 167, 172, 173, 180
Multiple Criteria Decision Making (MCDM), 146, 153, 157, 162, 167, 171, 175, 179, 180
Multiple Criteria Decision Support Systems (MCDSS), 146, 148, 176
Multivariate Analysis, 41, 42, 100, 171

N

Natural Language Processing Systems, 172
NFACT, Ii, 55, 56, 67
Nominal Group Technique, 158, 160
Nonmetric Variables, 42
Normative System Modeling, 160

O

Oblique Factor Rotation, 56, 73, 152
ORACLE, 163
Organizational Decision Making Rational Model, 157, 176
Organizational Process Model, 157, 176
Organizational Science, 146, 153, 167, 171, 172, 175, 176, 177, 179, 180
Organizationware, 161
Orthogonal Rotation Method, 56, 73, 152

P

Personality, 180
PERT Network, 161
Political Science, 157, 180
Principal Components Analysis, 63, 152, 153
PROC FACTOR, 54, 56, 73, 152
Proc G3d, 126, 128, 129, 140
PROC PLOT, Iii, I, 113, 115, 117, 118
PROMAX, 56, 73, 152
Pseudo F Statistic, Iii, 95, 97
Pseudo T^2 Statistic, Iii, 95, 97
Psychology, 157, 180

R

Rational Model, 157, 176
Raw Cocitation Frequencies, 36, 39
Raw Cocitation Matrix, 38, 39, 45, 63, 152
Reference Disciplines (RD), 145, 146, 148, 149, 153, 157, 167, 171

Relational Algebra, 163
Relational Database Theory, 162
Retain Statement, 134
Rotated Factor Pattern, 43, 56, 70, 73
Rotation Method, 56, 152
Round-Robin Procedure, 160
Routing DSS, 146, 148

S

Satisficing Model, 157, 177
Scatter Plots, 6, 101, 113, 128, 131
SCREE Option, 57
Scree Test, 55, 57
Set Statement, 133
Silent Independent Voting, 160
Statistical Analysis Systems (SAS), 23, 45, 49, 78, 84, 85, 93, 101, 103, 106, 152, 154
Statistical Bibliography, 1, 7, 8, 9
Strategic Management, 158
Structured Modeling, 162
Success Factors, 160, 165

T

Transaction Processing Systems (TPS), 159

U

User Interface, 153, 167, 171, 175, 176, 180
User Participation, 160

V

VARIMAX, 57
VAXIS, 116, 117, 118, 126, 127, 140, 141, 142
VTOH, I, 115, 118, 120, 121

X

Xmacro.Sas, 78, 154

NAME INDEX

Ackoff, R.L., 39, 40, 93, 154, 155, 159
Alavi, M., 181
Alter, S.L., 159
Anderson, R.E., 45, 55, 57, 70, 73, 77, 87, 89, 101, 152
Anthony, R.L., 159
Applegate, L.M., 39, 40, 154, 162
Baker, D.R., 9, 14
Bariff, M.L., 163, 164, 178
Bayer, A.E., 9
Bellardo, T., 11, 150
Benbasat, I., 154, 155, 161, 164, 172
Blanning, R.W., 39, 154, 162, 163, 172
Bonczek, R.H., 35, 37, 38, 39, 47, 154, 163, 172
Bui, T.X., 162
Carlson, E.D., 154, 155, 159, 172
Cattell, R.B., 57
Chang, A-M., 163
Chervany, N.L., 164
Courtney, J.F., 160, 165
Crane, D., 8
Culnan, M.J., 12, 24, 147, 148, 150, 175
Cushing, B.E., 148
Cyert, R.M., 157, 176
Davis, G.B., 175
Davis, R., 171, 172, 175
Delbecq, A.L., 158, 160
Dennis, A.R., 161, 162
DeSanctis, G., 161, 164
Dexter, A.S., 164
Dickson, G.W., 161, 164
Dolk, D.R., 162, 163
Dyer, J.S., 157
Elam, J.J. 163
Eom, S.B., i, 3, 8, 28, 36, 60, 73, 100, 145, 148, 146, 147, 149, 172

Everitt, B.S., 77
Farhoomand, A.F., 146, 147, 148
Farris, R., 145
Feinberg, A., 158
Gallupe, R.B., 161, 162, 172
Garfield, E., 1, 13
Geoffrion, A.M., 158, 162, 172
George, J.F., 162, 172
Ginzberg, M.J., 160, 178
Gorry, G.A., 146, 159
Gustafson, D.H., 160
Hair, J.F., Jr., 45, 55, 57, 70, 73, 77, 87, 89, 101, 152
Hamer, R.M., 102, 112, 152
Hatcher, L., 63
Henderson, J.C., 163
Hiltz, S.R., 154, 160
Holsapple, C.W., 163, 172
Huber, G.P., 161, 164, 176
Hulme, E.W., 1, 7
Jarke, M.A., 158, 162, 172
Jarvenpaa, S.L., 154, 155, 164, 172
Jelassi, M.T., 158, 162
Joachimsthaler, E.A., 181
Keen, P.G.W., 146, 147, 154, 155, 158, 159, 172, 180
Keeney, R.L., 157, 172
Kersnick, D.A., 164
Khattree, R., 95
Kim, J., 45
King, W., 160
Konsynski, B.R., 162
Kottemann, J.E., 163
Kraemer, K.L., 161
Kruskal, J.B., 101, 102, 103, 109, 112
Kuhn, T.S., 181
Kuo, A., 78, 81
Lanning, S., 154
Lee, S.M., i, 149

Lindsey, D., 14
Lucas, H.C., Jr., 164
Lusk, E.J., 163, 164
March, J.G., 155, 157, 176
Mason, R.O., 158, 163, 178
McCain, K.W., 23, 24, 28, 30, 37, 151
LcLaughlin, G.W., 9
Mueller, C.W., 45
Miller, D., 163
Mintzberg, H., 172, 176, 178
Mitroff, I.I., 158, 163, 178
Naik, D.N., 95
Newell, A., 172, 179
Nunamaker, J.F., Jr., 162
Olson, M.H., 175
Prichard, A., 7
Raiffa, H., 157
Raisig, L.M., 8
Raisinghani, D., 176
Ramsay, J.O., 102, 152
Robey, D., 154, 155
Rockart, J.F., 154
Rodriguez, J.I., 160
Rosengren, K.E., 7, 8
Sanders, G.L., 154, 155, 160, 165
Scott Morton, M.S., 159, 172, 180

Shortliffe, E., 171
Simon, H.A., 157, 159, 172, 177, 178, 179
Smart, J.C., 9
Smith, L.C., 11, 13, 14, 150
Sprague, R.H., Jr., 159, 172
Stohr, E.A., 154, 155, 158
Tan, J.K.H., 164, 165
Tatham, R. l., 45, 55, 57, 70, 73, 77, 87, 89, 101, 152
Theoret, A., 176
Turoff, M., 154, 160
Tversky, A., 172
Van de Ven, A.H., 158, 160
Wagner, G.R., 160
Watson, H.J. 154, 159
Whinston, A.B., 39, 40, 163, 172
White, H.D., 7, 11
William, S.E., 160, 172
Wish, M., 101, 102, 103, 109, 112
Young, F.W., 102, 112, 152
Zmud, R.W., 154, 155